Newton, Maxwell, Marx

OTHER BOOKS BY THOMAS K. SIMPSON

Maxwell on the Electromagnetic Field: A Guided Study
(Rutgers University Press, 1998)

*Figures of Thought: A Literary Appreciation of
Maxwell's* Treatise on Electricity and Magnetism
(Green Lion Press, 2006)

Maxwell's Mathematical Rhetoric: Rethinking the Treatise on
Electricity and Magnetism
(Green Lion Press, 2010)

FROM THE REVIEWS

Figures of Thought

Although Thomas K. Simpson's *Figures of Thought* is modestly
billed as an "appreciation" of James Clerk Maxwell's *Treatise on
Electricity and Magnetism* (1873), it is really a tour-de-force reading, one that lavishes on the *Treatise* the sort of scrupulous formal
attention usually reserved for the masterpieces of Western literature. Simpson thinks Maxwell's *Treatise*, in fact, *is* literature—and
he wants you to see it that way too. ... [His] genius is that, by the
end of the book, we believe him.

Diane Greco Josefowicz, *Isis*

Maxwell's Mathematical Rhetoric

Here we have a very lucid study of the cornerstone of electromagnetic field theory, that is, Maxwell's *Treatise on Electricity and
Magnetism*, not from the standard viewpoint of physicists, but
rather by looking at it as a piece of scientific literature, a work
of natural philosophy. ... The present book then has the merit of
showing not only how the concept of *field* entered into physics,
but also how important the choice of the mathematical language
is while introducing such a new physical concept. "I think it is
quite possible that we may yet be able to learn from Maxwell
something about the right way to question nature, and the form
in which the answers might come."

Salvatore Esposito, *Mathematical Reviews*

Newton, Maxwell, Marx

Spirit, Freedom, and the Scientific Vision

~

Thomas K. Simpson

Green Lion Press
Santa Fe, New Mexico

Manufactured in the United States of America

Published by Green Lion Press, Santa Fe, New Mexico, USA

www.greenlion.com

Green Lion Press books are printed on fine quality acid-free paper of high opacity. Both softbound and clothbound editions feature bindings sewn in signatures. Sewn signature binding allows our books to open securely and lie flat. Pages do not loosen or fall out and bindings do not split under heavy use by students and researchers. Clothbound editions meet the guidelines for permanence and durability of the Committee on Production Guidelines for Book Longevity of the Council on Library Resources. The paper used in all Green Lion Press books meets the minimum requirements of American National Standard for Information Sciences— Permanence of Paper for Printed Library Materials, ANSI Z39.48-1984.

Set in 11-point ITC New Baskerville and 14-point ITC Stone Sans. Printed and bound by Sheridan Books, Chelsea, Michigan.

Cover design by William H. Donahue.

Cataloguing-in-publication data:
Simpson, Thomas King
Newton, Maxwell, Marx: spirit, freedom, and the scientific vision / by Thomas King Simpson

Includes index, biographical notes

ISBN-13: 978-1888009-37-8 (sewn softcover binding)

1. Newton, Isaac. 2. Maxwell, James Clerk. 3. Marx, Karl.
4. History of science. I. Thomas King Simpson (1924–). II. Title

Library of Congress Control Number: 2011944624

Contents

The Green Lion's Preface

The three essays upon which the present volume is based were originally published over twenty years ago in the Encyclopædia Britannica's *The Great Ideas Today*. Although these annual supplements to the *Great Books of the Western World* were conceived as lasting contributions to the original set of works, the yearbook format seems to have suggested a more ephemeral role. Concerned that these insightful and timeless essays had not received the attention they deserved, Green Lion Press proposed to Dr. Simpson that they be given a book of their own. Intrigued by the possibility, Simpson began to consider how the essays, conceived independently of each other, elaborated a common theme. Inspired by new insights, he added the introductory and concluding essays and the connecting material that enunciate an encompassing vision of what science as a whole is, and what its still unrealized potential is.

The result is thus a new work, incorporating three essays in their original form, but extending them to constitute a new and cohesive whole. The Green Lion was delighted. We had originally wanted only to save from oblivion the original essays, which on their own are of timeless significance. They exemplify our aim as publishers to show serious readers the importance, the deep interest, and the sheer fun of returning to original sources, in the natural sciences no less than in the other humanities. They embody the lasting value that they bear as the fruit of Simpson's long contemplation, in both teaching and writing, of some of the most urgent issues confronting humanity. But now, with the new connecting and extending material, the three separate essays are shown to embody a new insight into how the fundamentals of science, rightly conceived, have always been connected to the highest and noblest aspirations of the human spirit.

The Green Lion expresses its gratitude to Encyclopædia Britannica and *Great Ideas Today* executive editor John Van Doren for permission to reproduce the original *Great Ideas Today* essays. For the convenience of readers, bibliographic references to volumes in *The Great Books of the Western World* series have been replaced in the present edition with references to standard editions.

Dana Densmore
William H. Donahue
for the Green Lion Press

Note on Reference Style

References to *Encyclopædia Britannica* publications appear as follows:

GBWW refers to the first edition (1952) of *The Great Books of the Western World*, followed by volume number.

GIT refers to *The Great Ideas Today*, followed by year and page number.

Preface

Preface

Introduction

Newton, Maxwell, Marx: three pillars of our western intellectual inheritance, yet each more honored today in encyclopedias and histories than read. What else do these three thinkers have in common, which brings them together here in a single book? Each of them, I believe, harbors a vision of *science*—not, however, as separated from other human activities but in thoroughgoing relation with our thinking, our moral actions, our productive powers, and with the general unfolding of the human spirit.

If we merely pay formal respect to these surprising thinkers without actually reading them—and reading them, moreover, with open minds—we are making a fundamental mistake. Not only did they often not mean quite what we commonly suppose they meant, but their thought is richer than we imagine, and in fact they have light to throw on our present perplexities. In short, they do not belong simply to the past, nor do they fit neatly into the categories into which we all too readily place them. We have much to gain by taking the opposite approach, and assuming that these authors wrote for the future—wrote, in effect, *for us*. Though we read them today in new contexts, and their language may seem surprising or strange, their ideas can be as fresh now as they were then.

Three great works stand out among their many writings: Newton's *Principia,* Maxwell's *Treatise on Electricity and Magnetism,* and Marx's *Capital.* Throughout my years on the faculty of what has become known as the "great books" program at St. John's College in Annapolis, Maryland, and Santa Fe, New Mexico, I have read, re-read and discussed these three books

3

with ever-rewarding results. Some of my thoughts about them were captured in essays written for Encyclopædia Britannica's annual series, *The Great Ideas Today*, whose executive editor was John Van Doren, himself an alumnus of St. John's and one of its most perceptive friends. The three principal chapters of the present volume reproduce three of these studies, one on each of these books.*

What appears to me on reflection to underlie my fascination with the three essays reprinted in this volume is a restless sense, which has grown upon me over the years, of the sharp contradiction they reveal between what we as humans might be, and what we actually are, and do. Human reason, and an earnest will to shape a better world, seem to be held in some form of *bondage*; we make immense progress in the sciences and in technological organization, yet we seem blocked in our efforts to turn these abilities to higher uses. We must, it seems, in some way learn to think together more intelligently, and turn our new scientific skills to overcoming the hunger, disease, ignorance, and warfare that stalk the world today.

It is unacceptable that mindless habit or prejudice should continue to hold us in such a grip. We are evidently, it seems to me now, still only at the threshold of learning what it might mean to become fully *human*. Yet this is not a question of mere good will, even if we could summon that: we have everything yet to learn. Humanity itself is in the midst, indeed, of a process of evolution, advancing haltingly but surely in a pattern of development, which I will call *dialectical*, at once painful and exciting.

Our three authors are themselves engrossed in this process. In different ways, they share this sense of struggle against a

---

* The three essays in this volume were first published in the following issues of *The Great Ideas Today*: "Science As Mystery: A Speculative Reading of Newton's *Principia*," 1992, pp. 96–169. "Maxwell's *Treatise* and The Restoration of the Cosmos," 1986, pp. 218–267, "Toward A Reading of *Capital*," 1987, pp. 74–125.

limiting bondage, and each achieves a victory that marks an advance of human thought. Together, they offer a striking prospect, one which it is of the highest importance to recognize and pursue.

Although these essays were written over two decades ago, I have not attempted to make changes to correspond to new findings or to reflect shifts in the unfolding of my own interpretations. Rather, I have chosen to let them stand as they first appeared, as I undertake the interesting task of reading them anew, all three together. Before this ambitious enterprise begins, however, it seems prudent to have a brief look in advance at the overall project, as I now have come to see it, in the present Preface.

These three works, and even my essays about them, will quite properly tell quite different stories to different readers. Nonetheless, I have arrived at a strong sense of one overall story that links them, and which I find compelling. Because it addresses so directly certain present-day concerns, which I suspect many readers may share, I offer here this synopsis of the story I find flowing through these works, gathering form and power as it goes. Readers who agree may find this sketch helpful; others may take it as a challenge that whets the sharp edges of their own thoughts. The separate essays will continue to stand on their own, without interference by such views as I urge here. My greatest wish is that this volume will effect some living junction between the potent thoughts of these three great authors and the hopes and concerns of readers of the present day.

The term *dialectic* is central to my own thinking about these matters and key to the unity I now perceive linking the three essays. I will explain more fully in a moment my understanding of its origin in the *Dialogues* of Plato and the meaning I find it retains for our world today. It was Hegel who showed that this concept of Plato's, translated in terms of time, becomes a theory of the unfolding of world history. Following this lead, I see each historical stage as beginning with some fresh affirmation or new insight, but subsequently passing through a phase of disillusionment, failure, or despair as this first phase reveals its dark face and inherent limitations. Yet then, in a third step, out of this darkness arises some genuinely new advance in consciousness incorporating the experience of these two earlier phases, while transcending them both. Through it all, nothing is left behind, and in this way our past intellectual history becomes incorporated both in a new sense of our present situation, and in our vision of a future prospect. We may speak, then, of an incorporation of past thoughts in the living body of our present experience; and this, it seems to me, is key to the vital interest of reading classic works today.

In writing about Newton, Maxwell, and Marx, therefore, I have tried to capture the vision with which each begins, the restless dissatisfaction out of which each in turn struggles, and the breakthrough each achieves in opening another in a succession of advancing prospects for the human mind. Their separate insights are facets of the complex, contradictory configuration of the human spirit in our own time; their struggles measure the depths and dimensions of the frustration many feel today. Most importantly, however, their successive visions of the possibilities of the human spirit add up historically to yield a sense of *prospect*, which, I am convinced, can give us strength.

Dialectic and Plato's Divided Line

The contradiction with which we live today was long ago prefigured by Plato—he is speaking of Athenians, but it as if he had looked ahead and seen us coming as well. We are prisoners, he says, bound head and foot in a cave lit by false lights; we are enchanted by a drama of shadows, projected before us on a wall. On its face Plato's fable is a simple image, though a telling one—but as Plato develops it in the *Republic*, it quickly gains richness and subtlety. Behind this picture lies Plato's larger vision of our human soul, or *psyche*. Psyche is the Greek word that Plato used; we still use it today.

The concept has taken many turns since Plato's time, and today the term psyche, when rendered as soul, may be awkward for some of us to use; it is hard to be sure just what it meant then, in the world before the arrival of the concept of God, and it is hardly easier to agree about its meaning today. Yet in many ways, whether we read its subject as soul or psyche, Plato's account touches us; he sets us thinking; and I suggest he may, almost by virtue of this very perspective which time brings with it, prove good physician to even our most modern needs.

Plato thinks of the psyche as partitioned functionally, the principal division being that between our intellect or *mind*, seat of our reasoning powers on the one hand, and the domain of *sensation* on the other. He proposes to image this in terms of a straight line, which we can think of as drawn vertically and divided into two un-equal parts (see Figure).

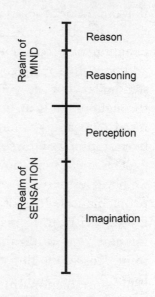

The lower, larger portion rep-resents the abundant realm of the senses, full of wonderful colors, tantalizing shapes, and enchanting sounds. Mind occupies the smaller, upper section. Each realm is divid-ed again: the upper portion of the realm of the senses is the domain of the *perception* of actual things, while the lower portion, again larger and more abundant, is that of *imagina-tion*—of stories, and the boundless picturing of things that may have been, or might yet be. The upper realm, that of mind, will be divided in a corresponding way, but we postpone for the moment the delicate question of the way in which that division of mind itself is to be understood.

Plato makes a most important point concerning the inner unity of the soul when he asks that we make these divisions in the manner of a musical tuning—so that each portion is divided in the same ratio in which the whole line itself has been divided. The whole will be, then, intensely self-similar, and harmonious. Mathematical ratios spell out as analogies, so we may say that the two inner realms look like each other, and each in turn looks like the whole: *mind is to the realm of the senses, as (within the realm of the senses) perception is to imagination.*

We go beyond the literal text which has come to us from Plato if we add, as we really must, that with the first division of the line, the *whole line is to the greater segment, as the greater segment is to the lesser*: the line then, and hence each portion of it as well, is divided in what the world has known from the time of Plato to that of Le Corbusier as the *golden section*, believed by some to be the most beautiful of proportions. With this addition, the soul becomes not only self-similar, but also most beautiful throughout. It is an image of harmony or peace, a vision of what the perfect tuning of the soul might be.

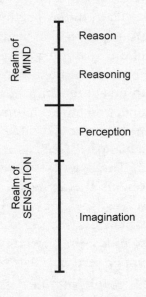

The next step, however, which we have postponed, may seem more problematical. It is of the greatest interest to Plato to make a further division, one within the domain of mind, corresponding to those we have already made of the lower division, and of the whole line itself. The lower portion of mind he calls *logos* (speech or discourse; reasoned argument or proof, or even, indeed, *ratio*, which is its Latin counterpart). We may think of this as the domain of the *discursive* role of mind: proving things in step-by-step arguments, or in the sequential stages of a process of calculation. I propose to term this *reasoning*, to suggest its step-by-step mode, and its character as process.

What is left, then, for an upper segment of the realm of mind, the highest division of the soul? Plato calls it *nous*, a term that gives us words like "noetic," but the question remains how *nous* should be translated. From its use in the dialogue, we realize that it means a direct insight into truth, and in particular access to the light of the *good*, unmediated by words or reasoning. Let us recognize its immediacy by translating *nous* very simply, as *reason*, suggesting a singleness of vision. This is in contrast to the extended term *reason-ing*, which we are using to distinguish the very different, ongoing and discursive process of *logos*.

It is reason in this very special sense that gives us the first truths, the underlying ones, on which syllogistic, linked arguments must be founded, or the first steps from which they proceed. It is akin to that vision we may get when, after passing through a long series of steps, we at last grasp the truth of a geometric demonstration as a whole, in a flash, and say "Oh! Now I *see*!" *Nous* sees things whole, and for it, the whole is primary; it is sometimes spoken of as *intellectual intuition*.

In using that phrase, however, we should be careful not to give the impression that it is merely "intellectual," for to Plato it is equally, and indeed more importantly, "moral": to see something as *true* is at once to see it as *good*. There is no tendency for Plato to see the mind as a calculating machine: we love *the good*, which is after all the ultimate source of all loves; we never think without caring.

Some moderns may not agree that such a realm of *reason* or direct intellectual insight exists, saying (as many do) that "all truth is relative," that what we call "truth" is merely a consequence of cultural upbringing and social habit—there were indeed plenty of Greeks, among them the Sophists, who said the same thing. If this were the case, the distinction we have been making would be lost, *nous* would have no role, and all the work of mind would come down to various forms of *reasoning*.

For Plato, however, *reason* is essential and the foundation of all else. If it were not so, all truth would indeed be only

relative or contingent, nothing truly *good*, and life would be
without meaning; and Plato does not believe this to be the
case. Evidently, his image of the soul presupposes a world of
the same kind to be its home. Moderns who doubt the existence of any such truth must doubt as well the existence of
such a world.

We might add a word about causality. How are we to answer
the question *why*? Aristotle and Plato do not agree in all things,
but they do agree about this: the ultimate answer to the question "Why?" is never in terms of pushes and pulls, but always
in terms of the goal or end of the action. For if it is true of the
world that the good is indeed primary, then the good is in turn
the true cause of all things; the Greek term for end or purpose
is *telos*, and so explanations in this mode are called "teleological." Other sorts of causes, such as pushes and pulls, are only
secondary—means to the end, or simply accidents. Wherever
and to whatever extent mind is involved, it is the *end* that is
causal, and what we want to know when we ask for an explanation of an action is, finally, its purpose. However, teleological
explanation is altogether out of favor in modern science; so if
we are to make a connection between Plato on the one hand,
and our three scientists on the other—as indeed we propose to
do—we have quite a course to run!

To hint at the justification of this excursion into such a curious theory of the soul, at the outset of an earnest discussion
of the sciences, let me simply propose that without that problematic top section and the truth to which it refers, neither
Newton's work, nor Maxwell's—nor even Marx's—would be of
more than historic interest to us today. It will be my proposition
that these authors are of serious interest to us today precisely
because there is indeed truth in the world, and they address it:
if we are in bondage today it is only because we find ourselves
cut off from that truth. We would have no sense of captivity,
subjugation, or servitude if we did not remember freedom.
Thus, the anguish of bondage of which I spoke first is, in these

terms, in effect a message from what Plato calls *nous*, and which we are calling *reason*.

Bondage and the Cave

It is Socrates who, in the *Republic* (the same dialogue in which the Divided Line is presented), tells the tale of the *cave*. Socrates asks us to picture prisoners (saying they are "like to us"). According to this tale, these prisoners—we ourselves—endure lifelong bondage, with our bodies, our very heads, so bound that we can look only toward a great wall before us. There, marvelous figures—we cannot know that they are only shadows—act out the hopes, delights, and triumphs, as well as the fears, disappointments and despairs that fill the courses of our lives. Light from fires behind our backs is casting these shadows, while the actors of this drama are slaves who walk back and forth—still behind our backs and unknown to us—carrying clay objects, which are the originals of the shadows by which we are so entranced. (It is hard to acknowledge that Plato had never entered a movie theater, had never seen TV!)

We prisoners never doubt that these shadows are in truth as beautiful and noble, as desirable or as terrible as they seem to us. They are the stuff of our hopes, of praise and blame, of our worst sufferings and our greatest rewards. We, Plato's readers, should not imagine this cave as dark or unpleasant, or that, for the most part, these prisoners have any sense of their captivity. The cave is the theater of our familiar lives. *It lacks, in short, for nothing except for truth.* The life of the prisoners lacks nothing, save freedom. How, indeed, could the prisoners know that their reality is mere shadow, their light false, their home no more than a niche hollowed out of a greater world?

They might, indeed, wonder whether anything abides (it does not, in the cave; all is transient), or whether life has any meaning beyond the momentary passing and vanishing of events (it does not, in the cave). Perhaps there might be

among the prisoners some who were distressed by such questions, suspecting that there might be, somewhere, something
abiding. It would be as if they were, as Socrates says elsewhere,
remembering something.

This is one of the great Platonic metaphors, and indeed,
before the *Republic* has concluded, we will have been shown in
a myth that everyone has already, in a realm prior to birth, seen
the great gallery of the true world beyond the cave. It is only
the *recollection* of that vision that causes us to doubt the cave,
and to feel that life within it is bondage. Of course, as myth it
is itself composed only of more shadows: since it cannot prove
anything, it folds in a marvelous way upon itself. This myth of
remembering can be effective only if it itself serves to remind:
it can be effective only if it prompts *nous* to see, and endorse,
its truths.

In terms of the Divided Line and the image of the soul, the
issue now becomes that of the doubtful uppermost segment.
If there were no such realm of truth, the cave would simply be
all. Reasoning, the calculative function belonging to the realm
of *logos,* would be thought to give scope enough for mind.
In the realm of shadows, there is endless theorizing, course-
giving and expertise concerning the phenomena. There are
recurrent patterns among the shadows that we can learn to
recognize and place our faith in. These faiths come to govern our lives. The modern sciences are generally understood
to be such areas of unbounded reason*ing*—they are thought,
indeed, to flourish precisely *because* of a liberation of reasoning
from constraint by any higher reason. Higher reason itself is
now regarded as prejudice and mocked as "metaphysics." We
have come to realize how transitory our theories are: history
has shown how they shift and change as new data are gathered.
The common understanding of the modern sciences thus constitutes an implicit denial of the existence of the fourth section
of the Divided Line.

As prisoners, how might we be empowered to grasp the
truth of the metaphor of imprisonment, to recognize ourselves

in those prisoners, and to cry out for release? Evidently, only that same faculty of reason within us could enable us to say "yes" to that tale, verifying it by the test of a light very different from that of the Cave. Reason is that wordless knower, recognizing a truth that neither words nor the evidence of the senses can assure.

Such truth inherently comes first, and while the argument in words and the evidence of the senses are of undeniable importance, yet these follow after, and their function is only to hint, not to tell. This, by the way, suggests the meaning of *education* in the Socratic sense: not to inform or simply convey—not to pour knowledge into an empty vessel—but to draw out what is already present, and to remind, acutely, of a truth already known. Since truth cannot be said in words, it comes in silence, and appears in the guise of *mystery*. Logos is powerless to account for it.

Since this recollection is of a world in total contradiction to the daily world of the cave and all its rewards, to become aware of one's bondage is to suffer acutely, and we perceive in it the seeds of tragedy. If we are to leave the cave it will be necessary first to be brought up short, to discover that one has been steering by a false light, and then to make a total reversal.

Within the Platonic dialogues the mode of this painful discovery is by way of questioning, and among them it is the *Gorgias* that perhaps most dramatically illustrates this process. Gorgias is by intent a good man who has achieved real success in the cave's terms: he is widely praised and highly effective as an expert in the newly formulated art of rhetoric, and when asked, he is happy to give an abundant account of this amazing instrument of human life. He enters on a dangerous path, however, when he accepts Socrates' invitation to say what his art is. He has no difficulty in heaping praise upon its powers, but as he is pressed over the course of the dialogue to penetrate to a proper definition, each effort by his own admission falls dismally short. He is at last confronted with the central question, whether the definition of this marvelous art must not include *justice*, a consideration that to this point he has altogether omitted.

The question appears never to have occurred to him; he is brought up short, in a terrible *impasse*—in Greek, the *aporia*, the point of no passage. He is led, before an embarrassingly public gathering, to a most reluctant recognition of failure, and an acknowledgement that his art, however powerful and effective, has been nothing worth. It is a failure, however, which at the same time incorporates a greater success, as he has glimpsed, however briefly, the light of justice beyond the darkness in which he has lived. Dialectic, of which Socrates is perhaps the ultimate master, has awakened the light of *nous* in Gorgias's own mind. He has been led by the contradiction in his own account of himself to remember something he had forgotten all his life.

We now have before us the pattern of what Plato has taught us to call *dialectic*, and which he proposes as the paradigm of all serious learning. It is a threefold process that presupposes a structure of the soul of the sort imaged by the Divided Line, and a life like that figured in the bondage of the cave. It begins with a positive thrust, as Gorgias at first undertook with such confidence in the praise of his own art. This is followed by a sharp negation, by downfall and the despair of *aporia*. Finally, out of this darkness emerges an altogether new *recognition*, the third phase of the trilogy of dialectic. It is only then that light from the fourth (highest) part of the Divided Line makes possible that advance toward new learning that the overall process is to achieve.

On the Greek stage, as exemplified in the *Oresteia* of Aeschylus, the three plays of the trilogy reenact this timeless threefold pattern of pride, downfall, and the dawn of some new, previously unimagined prospect—in Aeschylus's case, the birth of a new concept of justice in Athens. For Plato, as we have suggested, this dialectic is the pattern of all serious learning. We shall have to see to what extent it may open the way to a richer understanding of the long learning process that marks the unfolding of the modern sciences.

Dialectic Enters History

It is precisely at this point that we moderns have something new to say, some news for Plato. The unalterable stasis of the human condition that was the ground of ancient tragedy, as well as of the philosophy of Plato, lay in a timeless, cyclic repetition of the processes we have just described. An individual, even a whole people, might learn—might gain release to some degree from the cave—but then the inescapable cycle would repeat. We might say that *time,* in our modern sense, had not been invented, for nothing essentially changed, nothing genuinely new could arise.

The counterpart to this within the soul or psyche was a certain ultimate powerlessness in the form of what the Greeks called *ananke*, necessity. The faculty of *nous*, which we are here calling *reason,* could go only so far.

Human powers were not only limited in this way, but unequally distributed among mankind: we were supposed alike in being human, indeed, but far from equal in our capacities or worth. The conviction to which we subscribe today, that *all men are created equal* (with an asterisk now to the effect that *men* here is acknowledged to include women) was unknown—indeed would have seemed patently untrue or absurd—to Plato. In his eyes, some persons are clearly worth more than others, though it may often be unclear whether, in actual Athenian society, greatest credit is in fact given to persons of greatest worth.

The difference is the Biblical creation myth, which we share in myriad ways whether we are believers or not. This world view has the effect of granting absolute importance, by way of divine attention, to each individual person—something that has no counterpart in Platonic thought. As a consequence, there is in principle no limit to the possibilities open to any individual soul, and no unshakeable confinement, either, to the tragic repetition of the errors and limitations of earlier generations. With the concept of God as an omnipotent Creator, all things are possible, and thus a new order of hope arises, both for the individual person and for humanity as a whole.

Today our culture, our institutions and our philosophies are imbued with this new view, in which we are all caught up, whether or not we accept the theological doctrine out of which it has sprung.

The notion of *history* as an unfolding story, of *science* as advancing knowledge, and of *progress* as the hallmark of modern society—all are facets of the conviction that there is no need to accept the failures of the past as limiting the prospects for the future of mankind. Even those who are most skeptical about the achievements of what is today taken to be progress are likely to be reformers, people who have not at all abandoned hope but who simply hold to a different agenda.

All this is reflected in our concept of the soul and, as we shall see, in our concepts of reason and dialectic as well. The soul's powers have become similarly boundless, its strivings transcendent, and its highest work not merely to *know*, but to be itself *creative*. Scientists today sense that they have a part in shaping a future which, whatever form it may take, will be something of a sort the world has never before seen.

This new view of the human soul deserves a different name; I will call this active, creative soul *spirit*. Interestingly, it will be as appropriate to employ the term spirit in the context of Trotsky and Marx (in which it is directed to reshaping the world in which we live) as it is to employ it in connection with Augustine or Hegel (where it has explicit relation to the Holy Spirit of the Christian Scriptures—spirit as God in the world and within us). *Spirit* is thus the term we will now use for that highest faculty of the human soul; in the German of Hegel, the term is *Geist*, translated equally as mind or spirit. It is still the highest, directly intuitive reason of Plato, but with an entirely new aspect of striving toward the knowledge of something never before known, and the making of something utterly new on earth.

Earlier, we employed Plato's term dialectic to refer to that threefold process of learning that culminated in the intellectual insights of reason. Let us now follow Hegel in applying that same term to refer to an historical dialectic, a correspondingly threefold process that develops in historic time and culminates in insights altogether new to humanity.

It is dialectic in this modern sense which, I propose, links the works of our three authors, and despite the huge differences in the views they hold and the accounts they give, makes of their work one body of developing human thought. They speak deeply to one another even as they take seemingly opposed positions, and they move with an underlying directionality. We can see now (as they could not) that they are moving toward an ever richer understanding of the work of the human mind. To read them well may be to catch this spirit and to join in this ongoing process, which bears no less on our own circumstance than on theirs, and which, in important ways that we can barely recognize, entails consequences for our future.

It is important not to take this as simply a dialectic of *ideas*. Our three authors are not only putting forward new ideas of science and new tools of human understanding, but rather all are reshaping the world. This may be evident in the case of Marx, but it is no less true of Newton and Maxwell, who in a sense ushered in, or perhaps supremely captured, two successive ages of technology, the industrial revolution and the information era.

What had been a dialectic of thought, from which for Plato action had flowed only derivatively, now becomes incorporated into events as a dialectic of action. Marx will give this intimate involvement of thought and action a new name (albeit one out of the ancient dictionary), *praxis*. Whether we have Marx explicitly in view or not, however, it will increasingly be dialectic *as praxis* of which we speak as we track the works of Newton, Maxwell and Marx into the modern world.

Dialectic thus enters, and identifies with, history. What had been the cyclic Socratic conversation now has become a dialectical conversation between generations, and between ourselves, our past, and that future which lies always in our own hands. We have said enough, perhaps, to suggest how dialectic will be the thread we find binding our three authors to one another, and all of them to ourselves and to the anguish of our own time. We need not apologize for having lingered so long with Plato. In dialectic, earlier phases are incorporated in those that come after, and nothing is left behind. *Dialectic* is Plato's own term, and if the pattern of dialectic becomes our guide in following the conversation into the modern world, then as we go forward, we will always (as Scott Buchanan once observed) be meeting Plato coming back.

Newton

Newton, Maxwell, and Marx, read together, tell a compelling story of a long dialectical advance on the part of the human spirit. I have sketched the sense in which the term *dialectic* is being used here—in the modern form now, of a dialectic of history, though always with the older, root form from Plato's *Dialogues* in mind. His figure of the Divided Line, and his image of the Cave, will go with us.

Let us acknowledge that, like most stories, this one has no real beginning, at least not since Plato, Aristotle, and the ancient mathematicians and astronomers wrote. As we shall be seeing, our story has roots in the Biblical tradition as well. Newton builds on all of these foundations.

Our present acount must begin somewhat arbitrarily, with just a short reference to Descartes, who laid the immediate ground for Newton in a single step.

Descartes revealed the possibility of seeing the entire body of nature as one single system, altogether quantitative and mathematical. The works of nature were thereby rendered completely and thoroughly transparent to human intelligence—though strictly at the level of what we have been identifying as calculative reasoning, uncomplicated by ultimate issues of meaning or the good. It was thus a triumph of the mathematical mind, and appeared at the same time to be a welcome liberation from the mists and uncertainties of higher speculation.

Behind all the qualitative variety and color of the surface of nature there lay in truth only a featureless and undifferentiated quantity, and a single underlying reality, which Descartes called *extension*. To me this suggests that mysterious substance about which Euclid writes in the Fifth Book of his *Elements*, his account of ratio and proportion: he calls it merely *megethos*, *magnitude*. Here Euclid goes beyond both geometry and arithmetic, beyond figure and number, to deal only with the one continuum that underlies them all.

For Euclid, this is a purely intellectual object, but I suggest that, carried into three dimensions, this mathematical continuum becomes Descartes's *extension*. He endows it with real being, as a substance underlying all nature, and with this move he leaves us a world conceived as one single mathematical object.

Similarly, the *method* that Descartes exemplifies in his analytic geometry makes it possible to deal with this new world by means of a universal algebra. Every question can now be formulated as a solvable *problem* expressed in terms of an unknown quantity represented only by a single, utterly abstract symbol—that universal *"x"* which in a sense most characterizes our modern world. The clarity this brings to the work of the calculative mind is indeed breathtaking, and Newton, along with much of the western world, was clearly delighted by the prospect it opened. He early became a master of the new universal method.

The story of Newton's *Principia*, with which we are concerned here, however, begins with a very different aspect of his reaction to Descartes. Newton was as offended by this degradation of nature to the status of a merely quantitative object as he was pleased by the power the new vision offered: for Descartes had altogether excluded from nature any remnant of spirit or intelligence. The success of Descartes's natural philosophy lay exactly in the fact that nature was to run as a completely intelligible *machine*. There is no room for obscurity or mystery in the Cartesian machine. The world is a plenum—with no empty spaces—in which every portion acts directly by contact with its neighbor. This contact is manifest to the mind, if not to the eye, with simple and intelligible rules for the exchanges of motion upon their interactions. Actually, since this undifferentiated continuum is essentially a fluid, patterns of fluid motion such as vortices must play a major part; it is these which in Descartes's account serve to convey the planets in their courses.

The essential point is that no gaps arise in which mystery or doubt might hide—everything is explicit; all becomes clear. Yet this very gain of intelligibility, however great, might be seen as entailing a loss even greater. The reduction of nature to such a *machine* was as much an offense to Newton as the intelligibility it brought was a satisfaction to Descartes. Newton thus embarked on a strongly motivated, principled, and vigorous rejection of Descartes. It was this very pairing of affirmation and negation that served to generate Newton's *Principia*—a direct and dialectical polemic against Descartes.

Newton's own vision, which springs phoenix-like from this denial of Descartes, is a new, very different conception of a mathematical physics. Newton sharply separates what is *inert*—as all nature is for Descartes—from a second principle, which he sees as *active*. The first, inert, principle is matter; the second, the active principle, is force. Both principles are mathematically measurable, the one as the *mass*, or quantity of matter, the other as the accelerative quantity of force. The Laws of Motion express their relation.

Newton thus injects spaces into the Cartesian plenum, spaces in which a second principle, *force*, can operate. In a certain sense, the Cartesian intelligibility has been sacrificed. Nothing explains *how* forces work—the law of gravity, for example, which emerges from his study of the operation of the solar system, stands as a mystery. Newton does not undertake to tell us *how* the sun attracts the earth, but we do know that it does so according to a rational mathematical *law*—and all the workings of the solar system, and the moon and the tides as well, follow strictly from that law. The law of gravity is thus beautifully *intelligible*, though it is not mechanically *explainable*. For Newton, force, with the mathematical characterization of the law that it obeys, is an intelligible object, and thus belongs to the domain of spirit. It is clear as well, I believe, that Newton is on the track of that very special spirit in nature, the *vital force*. For Newton, animal motions, including those of our own neural systems, belong to the domain of spirit, and are *no less vital for being mathematical.*

This may be an important suggestion for us today, since we have come to the threshold of completing Newton's program. We can now envision giving a complete mathematical account of the animal system, including our own—brains and bodies alike. It will be very important for us to understand, if this is indeed true, that such an account does not necessarily describe us as machines, nor need such an account have anything *reductive* about it.

One clue to this important issue is the question of algebra. As we have seen, Descartes embraced algebra as the universal instrument of his new method; but Newton, though he uses it freely for practical purposes, very deliberately chooses to compose the *Principia* and the *Opticks* in the mode of *geometry*. His reasons for turning back to this tradition, rather than embracing the new mode, have been debated. I am convinced on balance that Newton vastly preferred geometry because the symbols of geometry—and unlike the universal x of algebra—bear *meaning*, and can be read and interpreted. In a nature full of spirit and intelligence, as Newton sees it, the mathematics that describes the natural world should be intelligible as well and should itself speak of spirit. I show in the essay on Newton how true this may be of the fundamental diagrams in the *Principia*. These diagrams both measure the acting force and point symbolically to that force's mathematical center.

Similarly for Newton all nature is, like the geometrical *Principia*, one intelligible text; it is the task of the scientist to *interpret* it. Causality operates in nature through and by means of force as agency; and the ultimate cause, Newton says, is always God. Newton's *Principia* and his alchemy are thus, taken together, instruments of his theology. Newton seeks this knowledge, however, not simply for its own sake, but in order to serve God better.

How would all this look in the very different terms of Plato's Divided Line? Let us think first of Descartes. It is very clear, I think, that Descartes sees nature and its study as a completely calculative enterprise: altogether a matter for the reason-*ing* faculty. If there is any role for the highest portion of the Line, it can only be to make the first principles of mathematics itself evident in full clarity. This is, I suggest, essentially the victory of the third section of the Divided Line over the fourth (highest) section. This victory cuts off from the study of natural philosophy the uppermost portion—reason, *nous,* or spirit.

To many, indeed, this seemed no loss, but rather a liberation from a realm traditionally embroiled in endless debate. But not so for Newton. For him natural philosophy is a realm rich in insight, one that itself has the power to bring man's mind closer to the mind of God. The *Principia,* in the System of the World, has lifted one corner of the veil, though God's larger plan remains for Newton tantalizingly beyond his reach.

Descartes thus sanitizes the world of science by cleansing it of the top segment of the Divided Line. What we see in Descartes is expressive of the modern world's great split in which our higher human concerns are banished from the domain of science. This has led us to a conception of science as an activity that is strictly delimited, sterile and, as we say today, "objective." With this has come, as we have witnessed, the headlong assimilation of the rest of scholarship to this same criterion of sterile objectivity. Nature, if it has indeed become a mere machine, may be of great technological interest, but hardly more than that. Divested of meaning, it is of no more than practical interest to us. However, spirit may in its turn rebel furiously at such an insult!

Can we now find a relation between Newton and the Divided Line? Newton's reaction is an outcry on behalf of that very spirit that Descartes had tried to exorcise. Newton's work appears highly spirited throughout: he has, indeed, a central program, and is single-mindedly on the track of the very intelligibility demanded by *reason*, the highest faculty of the soul. Initially, then, it seems that Newton is restoring the fourth section of the Divided Line that Descartes cut off.

And yet, to just what extent is the full intelligibility demanded by reason and spirit possible, or really even sought, in Newton's work?

If Newton insists that God as cause is the ultimate object in natural inquiry, this is hardly the answer sought by reason. For Newton, God and His purposes ultimately transcend our powers of reason, and in this sense the laws of nature are as inscrutable as any laws handed down on clay tablets. The laws themselves, then, are (as we have said) mathematical and intelligible objects for mind, but, beyond them, God's plan is not such an intelligible object. Our minds grasp the laws intellectually, but we are not privy to the purpose that lies behind them, as *nous* would demand.

The creation thus becomes a sacred text, and the pursuit of science, for Newton, is an aspect of the task of *interpretation* of God's word. We can interpret God's laws and study to carry out our tasks as responsible members of God's realm. We might say that Newton's model for science is essentially feudal: we are servants of a transcendentally wise and beneficent Lord. We cannot fathom God's ways, but we can comprehend the wonders of the effects they produce.

Neither Newton's science nor Descartes's, then, altogether affords satisfaction to the human spirit and the demand of reason for insight. Each opens brilliant new realms to the human mind but neither restores access to the uppermost section of the Platonic Divided Line.

As we turn to the work of James Clerk Maxwell, we shall see again the interplay of acceptance and rejection of the works of a predecessor. As Newton initially delighted in, and took much from, Descartes, so Maxwell mastered and learned from Newton. And, similarly, as Newton rejected Descartes, Maxwell thrust Newton aside to move forward in a new direction, toward a further and inherently better way of thinking about the natural world.

Maxwell

Newton's successors did not understand his project in the terms he had intended. The new world of the industrial revolution was rapidly building machines and bridges and pressing into new realms of physical phenomena. It welcomed the *Principia* as text for a new age but it wanted results. The metaphysics and theology lying behind the calculations faded from attention.

The concept of force as Newton's successors used it proved extremely useful, but retained nothing of Newton's sense of *spirit*. Indeed, in time it became quite possible for even well-informed scholars to use the oxymoron, *Newtonian mechanism*, losing the distinction between Newton and Descartes and quite possibly missing altogether the central point of Newton's life work! We must note also that friends and editors had mounted a deliberate and quite effective campaign to bury Newton's embarrassing work on topics such as alchemy and theology. Today, with these works now opened to us, we are privileged, in a way generations of readers before us have not been, to understand Newton's thought as a whole.

By the nineteenth century, however, tradition had effectively transformed Newton into a mechanist. Newton's laws were universally accepted because they worked. Primarily because they were required on examinations, they became something of a tyranny over the minds of generations of students. Indeed, in England mastery of the Newtonian mathematical physics had become virtually the sole path of entry into the practice of serious science, and at the same time had become the exclusive privilege of a social class able to afford the price of a university education.

Newton's work had thus lost the vigorous dialectical spirit with which it had begun. From the point of view of the Divided Line, all this had the effect of returning science to the formal enterprise it had been for Descartes—appropriate to the third realm, that of reasoning, but once again sealed off from any access to the realm of reason itself. Science, thus reduced and isolated, remained an implicit affront to the human spirit, which in the long run may never settle for less than the full scope of mind's inherent powers.

It was this view of a science reduced to mere mental discipline—challenging indeed, but sterile—that confronted James Clerk Maxwell as a student in the middle of the nineteenth century. This was especially true when he transferred from the liberal, relaxed, and reflective atmosphere of Edinburgh University to enter upon a highly rigorous and disciplined study of natural philosophy at Cambridge.

It will be important to appreciate what, besides intellectual brilliance, Maxwell brought with him to Cambridge, and therefore I shall say a little more about his rather special personal background—significant, I believe, for the form his science was to take. Maxwell, it is perhaps fair to say, remained at heart a Scotsman at a time when things Scottish were under severe stress from a far more powerful English culture and economy. He was always in some sense out of place at Cambridge, for all his skill at mastering the English ways, and despite his celebrated success as a professor there in later years. Yet Maxwell was no simple nationalist. His father was Scottish; his mother, who died when he was a child, was English; Maxwell's mind was rich enough to perceive and even embrace both cultures—and then in some way, the *whole*.

Along with the complexity of his dual Scottish and English heritage, Maxwell's formative years reflected both urban and rural influences. Though he was born in Edinburgh and belonged on his father's side to an Edinburgh family, he was raised on a family estate—which Maxwell loved, and on which he was later to do much of his best scientific work—in the depths of rural Galloway. As a schoolboy, however, he was abruptly transported from Galloway into the strictures of the new, very English Edinburgh Academy, where his rural ways quickly earned him the nickname "Dafty." It appears that throughout his life he typically spoke in Galloway accents the British had difficulty in understanding, and he tended to veil his thoughts in a mask of irony and indirection that his listeners, and at times his readers as well, had difficulty penetrating.

After distinguishing himself as a student in the Academy, he was sent not at first to England but to Edinburgh University, which at that time was still holding out firmly against the new English demands. There Maxwell studied for two happy years—highly significant for his later work—immersed in Scottish traditions of broad liberal and humanistic education, in many ways the opposite of those he would meet at Cambridge.

From this Edinburgh idyll, he was transported to Cambridge University, where, soon migrating to Newton's Trinity College, he distinguished himself in those mathematical rigors of which we have been speaking. In Scotland, the geometrical methods which had been so important to Newton, had been carried forward by such distinguished geometers as MacLaurin and Simson and remained a hallmark of the Scottish approach to a humane mathematics. In England, by contrast, skill in analysis was thought to be excellent mental preparation for future prelates and statesmen. Maxwell came to excel in both worlds—but I believe we will see that the complex stresses which ran through this entire experience shaped his mathematical soul, and had a great deal to do with the revolution he was to carry out in developing the concept of the electromagnetic field.

As a fellow at Trinity, following success in the examinations of the sort we have described, Maxwell immersed himself in the researches in electricity and magnetism of Michael Faraday, which had recently been gathered into the three volumes of *Experimental Researches in Electricity*. To read these works with care, as Maxwell did, was to embark on an immense pilgrimage into the mind and ways of a brilliant but remarkably unassuming person. Not many among the privileged in England would have chosen to follow, as Maxwell did, in Faraday's humble footsteps. Faraday was patronizingly praised as a "discoverer," but (so it was thought) anyone with a proper university education would understand that Faraday, innocent of mathematics and untaught in the formal sciences, could contribute only as a skillful laborer in his vineyard at the Royal Institution, while leaving serious science to those of the higher ranks.

Some reassertion of Maxwell's Galloway convictions must have led him, therefore, not only to follow Faraday faithfully through the intricate paths and byways of the *Experimental Researches*, but to devote himself to a systematic reformulation of Faraday's diagrams of the *lines of force* surrounding configurations of magnets, shaping new analytic forms to capture as nearly as possible Faraday's own insights. Faraday had announced himself to be, without apology, an "unmathematical philosopher," and had asked Maxwell in a letter whether it would not be possible for the results of mathematical researches to be expressed in language which persons, like himself unlearned in mathematics, might understand. In response, Maxwell composed the paper, "On Faraday's Lines of Force," as nearly as possible in an intuitive, geometrical mode accessible to the untrained mind; it was a gift to Faraday.

Maxwell's opting for Faraday and for geometry over analysis lies, I believe, close to the foundation of his convictions concerning natural philosophy and mathematics, and equally at the root of a revolution he carried out in formulating electromagnetic theory in field form. This claim may seem difficult to maintain, for admittedly Maxwell went on to perform brilliant work in the analytic mode. Nevertheless it is my suggestion that in Maxwell's field theory the analytic functions are shaped to capture as directly as possible the spirit of Faraday and geometry—shaped to speak intuitively to the mind's eye in ways which conventional analysis would not do. Writing on Faraday near the end of his own life, Maxwell held to a position he had announced at the outset, that among them all it was Faraday who was the real mathematician!

The nature of mathematics, and of mathematical physics, is seriously in question here. As Maxwell translated Faraday's insights into the symbols of formal mathematics, the geometrical and visual mode remained primary. In stark contrast to the practices of the "professed mathematicians" (Maxwell's phrase), the new analytic mathematics of the field, which Maxwell himself was developing, remained only instrumental, secondary to his guiding intent.

A further turn of Maxwell's thought occurred well into his scientific career, when he first encountered Lagrange's equations of motion. These characterize, in terms of the Continental concept of *energy* rather than Newtonian force, the motion of a *connected system as a whole*. Maxwell retained a deep concern for metaphysics, which had fascinated him at Edinburgh; for him this primacy of the whole was clearly a fundamental idea and a matter of first importance. He thus made certain in his *Treatise on Electricity and Magnetism* to place Lagrange's equations at the head of his reasoning and then to derive his own system of equations as a particular case of Lagrange's. Only by reasoning from these equations (that is, by systematically determining those values of their coefficients that would give form to the electromagnetic case) would he arrive at electromagnetism as an instance of the general idea of a connected system as a whole. In this way Maxwell's electromagnetic equations are, in his own presentation, infused from the outset with the spirit of wholeness appropriate to the concept of the field.

Maxwell's analytic theory ultimately takes the form of four symmetric differential equations, and these—today recast in relativistic terms—stand as foundation stones of modern mathematical physics. By speaking universally of relationships that obtain at every point in a space, a differential equation has the remarkable ability to describe the space as a whole.

By contrast, the individual bodies and the particular forces acting between them are for Maxwell only the last and most problematic concepts to emerge. Maxwell has in effect initiated an inversion in the structure of mathematical physics. The elementary bodies and their relationships, which were primary in Newton's theory, become secondary and derivative in Maxwell's. The whole, which would be derivative from Newton's theory, is the foundation of Maxwell's.

This inversion has sociological significance as well. Through the concept of the field, grasped by way of its patterns, Maxwell has made electromagnetism intuitively accessible to every inquiring mind. This new approach thus undercuts both the primacy of sheer mathematical symbols and the necessity of an aristocratic formal education with which to master them. This, in its effect, was Maxwell's great gift to Faraday, and even more largely his gift to what has been termed the *democratic intellect.**

We moderns are failing to catch the significance of the revolution Maxwell was undertaking in our behalf. We are faltering in our efforts today to comprehend whole systems in their entireties—whether environmental systems or the global human community—because science hasn't learned from Maxwell to reason in terms of the whole. The difficulty is aggravated because we have not incorporated Maxwell's democratic insights into the larger body of our social thought. Our society still treats mathematics and the sciences as matters reserved only for specialists, and hence abandons the opportunity to learn from masters such as Maxwell.

We should remember, however, that Maxwell's revolution is against the *distortion* of Newton, not against Newton himself.

---

* Davie, George Elder. *The Democratic Intellect: Scotland and Her Universities in the Nineteenth Century*. Edinburgh, The University Press (1961).

For all their evident differences, at a deeper level, concern for the unfolding human spirit unites them. Newton's notion of the human spirit is expressed in a paradigm of law and obedience, which we have characterized as essentially feudal. Maxwell reflects the larger course of human history when he carries this human spirit to a democratic manifestation.

Newton describes a layered world of command and obedience at a distance; Maxwell fills all the gaps. Each of his field equations applies everywhere equally; as a set, their solution takes the form of a statement about the whole. In this sense they become equations of a coherent, intelligible cosmos. They are, as well, the very image of a restored and democratic social whole. Things viewed in this way, in their global relationship and as constituting a whole, *make sense*—a sure sign that the way has been opened once again to *nous*, and to the full scope of reason and the human spirit.

Marx

To many readers it may seem strange to include an essay on Marx in a study dedicated to the unfolding image of modern science. But as the essay on *Capital* in this volume will spell out, Marx sets out a theory of capitalism in the very image of the *Principia*: as a closed mathematical system strictly governed by one overall law of force. Most critics seem to have missed this point—perhaps students of *Capital* have too seldom been students of the *Principia* as well—but it would be no exaggeration to say that to a remarkable extent Marx is the Newton of capitalism.

Admittedly, Marx's style is the very opposite of Newton's. Marx was a classical scholar, and he reasons rather more like Aristotle, whom he studied and appreciated, than like a mathematical physicist. Yet the resulting work is founded solidly on the definitions of strict mathematical quantities, and the resulting motions are traced as rigorously in their detail as the planets are tracked in *Principia*. Here, as in the *Principia,* the purpose goes far beyond developing a merely descriptive mathematical account, or even making accurate quantitative predictions. In both cases the objective is the same: to give a *causal* account, well founded and thorough, of the operation of a system governed by mathematical law. *Capital* is devoted to precisely that goal.

Explanation of this claim will take some time, and will be the burden of the essay on *Capital*. It may help, however, to offer a few hints in advance. For *mass* (or quantity of matter), the underlying measure of the *Principia*, Marx substitutes the *undifferentiated labor hour* as the precise unit of measure in capitalism. And for the driving law of capitalism as a social system, Marx sets out the *law of surplus value*. It is not a matter of moral choice, but a necessity of the very system, that every capitalist enterprise *must* run in such a way as to maximize profit, here rigorously defined in terms of surplus value as the underlying measure.

What sort of accomplishment does this achievement of Marx's represent? It is of course a crucial breakthrough on behalf of reason: to be able to grasp society as an intelligible system is of the highest importance to the human spirit, and already a step toward freedom. Yet Marx's analysis reveals this society itself to be unfree: run by laws of strict necessity, yielding results over which reason in the higher sense has no control, and conditioning consciousness itself. It is an even more remarkable accomplishment that the author is able to step outside this system to bring to the work the fullest play of reason.

In this sense, *Capital* is perhaps a new kind of literature. It is the product of a human mind that on the one hand develops the theory and rigorously traces consequences of the system and on the other judges from a higher point of view the human costs and affront to human dignity that each stage of the process entails. This judgment is not a vain or idle outcry, but a reflection of reason itself.

Marx writes in a mode of complex irony. He is scrupu-
lously painting a precise portrait, yet he despises the picture
that emerges and does not spare the reader any of its pain or
ugliness.

If *Capital* were only as we have so far described it, it would
be a work of darkness and despair. In fact, however, it is the
opposite: out of the contradiction emerges, dialectically, a
vision of something altogether new. What Marx is seeing is that
capitalism works despite itself to generate within its bosom a
system of an entirely new order.

Painstakingly, Marx outlines how this process works. The
enterprises that capitalism fosters are purely individual and
separate. They work according to the law of surplus value
toward ends of their own. They tend by necessity to aggregate
ever greater blocks of capital in large industrial operations.
They replace the skilled workers of earlier, small enterprises
with organized, analyzed, mechanized and more efficient
modes of production in which the workers become only cogs
in a gigantic enterprise. Workers are thus inevitably drawn into
a homogeneous body of common humanity in which their
individual skills and efforts are eclipsed by the magnitude of
corporations. They are hired and fired at the convenience
of the enterprise, and they move from the tendance of one
machine to that of another as the evolving enterprises demand.

Marx's prescience seems almost uncanny. One marvels at
the extent to which Marx, though he is looking at a nineteenth-
century world, is actually foreseeing what we now know as the
global economy. He is not only foreseeing but analyzing what he
sees in detail. And we recognize that he is articulating today's
growing sense that humanity shares one common destiny.

Capital operates outside praise or blame, and the players
in the capitalist system, whether they be bankers or laborers,
function as they must. Any personal attempt to opt out would
leave the mechanism itself essentially unchanged.

Marx by no means disparages the emergence of this modern world: disparagement would in any case be merely idle

activity. Far more interestingly, he sees capitalism developing amazing new technologies capable of transforming the very character of the human scene. Everywhere the expansion of capitalism has entailed the development of rational planning on the immense scale of the global corporation, and with it the cooperative organization of the labor process to the achievement of ends that would have been totally beyond attainment by individuals or smaller enterprises.

At the same time, however, mankind has become increasingly the servant of these new global enterprises, which run only to achieve their own ends. Marx says this in the midst of the early industrial revolution; everything he says seems to ring only more true today.

There is no question that humanity can function intelligently for common purposes on a large scale—this is being widely demonstrated on a daily basis. What is at issue is the matter of actually turning our new powers away from strife and competition. Each individual human being would, in Marx's vision, gain real access to the new resources which surround us today, and thus be enabled to act freely—his term is *spontaneously*—in accordance with quite individual and personal aims.

Capital is thus not only a seriously scientific work in the Newtonian sense, but one that appraises its findings according to the highest reason, devoted to justice and the highest human good. Marx is evidently proposing a fundamental extension of the concept of science itself, to include *critical* reason.

Capital is not a rhetorical work, summoning to action, since the science itself foretells the collapse of capitalism with or without action. Politically revolutionary movements have indeed sprung from it. But some have misunderstood the science and missed or sacrificed Marx's vision of the flowering of human life. The entire era of Stalinist communism, for example, constituted in many respects a complete misreading and a travesty of Marx's intention, which is everywhere devoted to liberation of mankind, first and finally, as free individuals. Marx's goal is liberation from the constraints of a mechanical system. The test of liberation is the possibility of spontaneous human action. Marx foresees that modern methods could easily make such free human action possible, were they devoted to that end.

In the third volume of *Capital*, Marx takes criticism to a higher level. He asks, in effect, how is it that mind can become sealed in a system from which it seems powerless to escape? He traces the way in which capitalism paints its own picture of the world in which we live: a complete system of social relations, institutions, systems of thought and the vocabulary of daily life. Criticism must therefore penetrate very deep, through all these structures of thought, language and society, which form the very air we breathe.

Capital must in this way look critically at its very self, its own methods and its own language. Marx's project is intensely dialectical, invested everywhere with the white heat of contradiction. Yet it is aware always of the ultimate autonomy of free reason which creates the work itself, and it invites that same free act on the part of the reader—with whom it is communicating through so many veils.

The result is a level of a relentless wholeness of view that is exactly and always the demand of that highest reason which Plato called *nous*—now, however, carried into history and housed, as we have seen, in the vessel of science. This is the same highest reason with which Newton thrust forth beyond the limitations of Descartes, and the same reason by which Maxwell in turn forged the concept of the *field* to escape the limitations of Newton. Reason wants the whole answer in complete freedom: it will not rest with arbitrary pronouncements or mechanical limitations. Thus, complex as the story has turned out to be, I see through these three works one long struggle of the free human spirit—Plato's *nous* committed to a dialogue that evolves through human history. This story comes now to us, as readers of all three, inviting us to understand and carry the project forward in our own time.

Newton

Introduction to the Newton Essay

OUR FIRST ESSAY is *Science as Mystery*: *A Speculative Reading of Newton's Principia*. By 1992, when *Science as Mystery* was first published, it had become evident to historians of science that Newton was very seriously concerned with other areas of thought than the one that was the overt subject of the *Principia*, which he called "natural philosophy," and which we now call "physics." I was especially caught by the realization that Newton worked regularly and over a long period of time on *alchemy*, and the essay thus takes that as its theme, proposing an *alchemical reading* of the *Principia*.

In the years since *Science as Mystery* was written, it has become even clearer that Newton worked with essentially equal effort and conviction in other domains. We may count four principal areas of intense interest to him: (1) what we would recognize as *science*, in the manner of the conventional reading of the *Principia* and the *Optics*; (2) *alchemy*, in both its theoretical and its practical phases; (3) *theology*, a subject on which Newton wrote intensely and extensively; and (4) *history* or *chronology*, which he evidently regarded as a subject of importance co-equal with the others.

What is so impressive today is the realization that all of this was the work of one single, encompassing mind: and the recognition to which this leads, that these are, for Newton, deeply related domains within one single truth. *Truth*, for Newton, *is one*. Work in each domain, being part of one single whole, inherently relates to all the others. *Science* is not, then, an isolated or separable domain of truth, and even the mathematical physics of the *Principia* bears meaning for Newton of a sort we no longer expect, or hope for, in our sciences today. A fresh reading of the *Principia*, then, may come as a welcome message of hope for our jaded minds: if Newton is right, the sciences—without ceasing to be all that we expect in terms of discipline and rigor—may be *meaningful* in ways we have ceased to imagine. His lesson might well be as applicable to quantum mechanics or relativity theory today, as to his own System of the World.

How all this relates to an *alchemical* reading of the *Principia*, we leave to the essay to show.[1]

---

[1] References to the *Principia* are to the following edition: Cajori, Florian, ed., *Newton's Principia, Motte's translation revised* (Berkeley, California: University of California Press, 1960)

Science as Mystery: A Speculative Reading of Newton's *Principia*

I S IT POSSIBLE that the workings of certain human minds are so ceaseless and boundless that they eventuate in books too big to be read? Something of this sort seems to be the case with Newton's *Principia*—more fully, the *Philosophiae Naturalis Principia Mathematica* (*Mathematical Principles of Natural Philosophy*). Not merely the daunting task of mastering its mathematical demonstrations, but the challenge of incorporating into one's own thought its hierarchy of stacked purposes and multiple lines of implication has over the years perplexed even the best-equipped minds. Those with the technical ability to deal with the mathematics have tended to be pressed to other tasks or have missed the metaphysics, while most readers have been able to get only limited distances into the text. Within the work itself, though rather late in the game, Newton acknowledges the difficulty; at the beginning of Book III, he roundly advises against having read Book II and prescribes a road map for Book I which unfortunately would stop well short of some of the most essential material.[1]

Such maps have disserved readers over the years; but even stalwarts who make their way to the end of the first book normally leave the second book unopened and thus get only a very partial sense of the terrain as a whole.

The result is that it has been the habit of scholars and general readers alike to assume a common understanding of

---

[1] Newton's advice comes at the outset of the book (*Principia*, tr. Motte-Cajori, Book III, p. 397) in these terms:

> ...not that I would advise anyone to the previous study of every Proposition of those Books; for they abound with such as might cost too much time, even to readers of good mathematical learning.

what the book was about. It has been recognized as a portal
through which Western society entered upon its modern era—
a portal rather too confidently characterized as the "Scientific
Revolution." Newton, it has been thought, set in the book a
magnificent example of something we call the "scientific
method," and he sealed the world's faith in it by exhibiting
its ability to compute and predict the motions of the heavenly
bodies. Its basic terms, *mass* and *force*—even *time* and *space*—
have become so current among us that, encrusted as verbal
and mental habits, they have become almost impenetrable to
critical reflection.

Just now, however, we may be in possession of a historical
opportunity to make a new start at these matters. We have
reached a point in our relation with the sciences at which a
combination of the unsettling of fundamental concepts, on
the one hand, and disenchantment with practical outcomes,
on the other, is causing us to wonder about the sciences and
to question the concept of progress with which they for so
long lured us. This is a moment, then, at which it is timely
to reconsider our understanding of the idea of science itself.
Concurrently, new evidence has been coming in that we may
have been misunderstanding the *Principia,* our presumed
guidebook to the modern sciences, as well. Newton certainly
did think he was precipitating a revolution in thought, but his
vision of the nature of that revolution may not have been much
like that which we have imputed to him.

Close reading of even a limited assignment in the text
has always suggested that we and Newton might be looking
at things differently. There are strong indications within the
text that he was on the track of *science* in something closer to
Aristotle's own meaning of that term, as the theory of things
true, and known with the greatest certainty: systematic induction was for Newton, his text suggests, a refined method for
making sure of absolute truth. We, on the other hand, posi-
tivists all in these matters, had in effect to read his intentions
out of the text in order to take it in our own mode as a primer
for our own understanding of scientific method, essentially

a method for evading the really interesting questions. The general excuse for this performance was that Newton, one foot in the future but one equally somehow stuck in the past, wrought better than he knew. History had kindly edited his text to make it say what future generations needed to derive from it.

Now, however, revelations of current scholarship make certain of Newton's own intentions inescapable and may rightly send us back to the *Principia* to give it a less cavalier, more speculative reading. Quantities of Newtonian manuscripts are coming to light and receiving competent scholarly interpretation, which begin to take fuller measure of his work as a passionate theologian, an obsessive historian, and above all, for our present purposes, as a magnificent alchemist.[2]

---

[2] Many fascinating accounts of Newton's alchemy are now appearing. The most useful general introduction is probably by Betty Jo Dobbs, who has written numerous interesting articles on this subject as well: Betty Jo Teeter Dobbs, *The Foundations of Newton's Alchemy, or "The Hunting of the Greene Lyon."* A striking overview by David Castillejo of the weaving of the many strands of Newton's thought, based on extensive study of the new manuscripts, has been particularly valuable in the preparation of the present essay: David Castillejo, *The Expanding Force in Newton's Cosmos.* The first chapter of that work is devoted to a review of the alchemy.

In general, studies by Richard Westfall have led the way in much of the current scholarship. His biography of Newton, *Never at Rest*, which includes an extended bibliographic review, and his earlier work, *Force in Newton's Physics*, are good starting points. An earlier biography by More remains useful: Louis Trenchard More, *Isaac Newton. A Biography.* The classic biography is that of David Brewster, *Memoirs of the Life, Writings, and Discoveries of Sir Isaac Newton.* Excellent editions of Newton's mathematical papers and of his correspondence are now available, as well as editions of his unpublished papers. These, together with a variorum edition of the *Principia* itself, provide a rich prospect for the interested reader. A number of very interesting collections of essays have appeared in recent years. Most especially, the reader is urged to refer to the essay by Curtis Wilson, "Newton's Path to the *Principia*," which appeared recently in the present series (*GIT* 1985, pp. 178–229).

These disturbing truths about our author are not indeed altogether new. Dark reports had always had it that Newton engaged in covert projects in alchemy, and it was always suspected that he had a twisted bent for synoptic history, theology, and the interpretation of Scripture, especially of the prophecies of Daniel and John. Indeed, it has long been known that he had indulged in alchemy, though the cover story was that he had only copied and recopied, presumably in fits of nervous compulsion, the manuscripts of others. Now a deluge of his hidden and despised manuscripts has burst upon the scholarly world, by way of what must be one of the more embarrassing archival misdemeanors of modern times—an auction at Sotheby's in which a wealth of Newton manuscript materials was dispersed to the winds of the marketplace, with only the Sotheby catalog left to document their existence for posterity.[3]

Most of these have since been tracked down, and enough have been read and reported to dispel any notion that Newton was simply mad, or a part-timer in these exotic fields. No; it is now quite clear that the same brilliant and relentless operations of mind which had generated the *Principia* and the *Opticks* went at least equally and concurrently into what we can only call the most earnest commitment to alchemy, and an immense, disciplined study of history and theology as well, the latter based upon a critical evaluation of scriptural text which anticipates in method and concern our modern biblical textual criticism. Prospero himself has emerged before our astonished eyes, with the *Principia* his magic book.

To take the hardest point first: we must face squarely the fact that Newton was a full-scale professional alchemist. One scholar in a position to make an estimate suggests that never before or

---

[3] It was the collection of Lord Keynes which was placed on the block in 1936; as a result, it was fragmented into over one hundred auctioned lots which went to some thirty-two different buyers (Castillejo, op. cit., pp. 13, 120; Westfall, *Never at Rest,* pp. 875ff.). Over the years since, scholars have tracked down and published at least the locations of most of the components of this diaspora.

since has there been such a master of alchemy—both of the
intricate texts and of the solemn and exacting practices at the
furnace. Suddenly, a sense of mystery falls over the *Principia*, once
the very key to the liberation of the modern world from all such
dark suggestions. Thus, I propose that at this point, at which we
have reason in any case to reconsider the nature of the sciences
themselves as the foundation of our dubious modern era, we
take occasion as well to return to Newton's text and embark
upon a more speculative reading, with the Newtonian alchemy
in mind as a leading thread. The *Principia* is a deeply dialectical
work, meant to advance one thesis and to dispel another. Given
that terms such as *Newtonian mechanism* are so often encountered today in summaries written at a second remove, it may be
surprising to discover that the *Principia* is thoroughly organized
as a polemic *against* mechanism. In its place, Newton invites, I
believe, a legitimate sense of science as mystery. Newton's term
force, a surrogate for the alchemist's *spirit*, is essential to the
operations of nature once clockwork mechanisms have been dispersed. It is meant to invoke a mystery, to make way for agencies
in the cosmos which are not material, and so to leave room for
the divine. Newton's theology is thus by no means incidental to
the rest of his work: it is even possible that Newton understands
the *Principia* to hold a crucial position in the prophetic succession, as a new and decisive revelation of God's work, and that
thus a divine voice animates its propositions.[4]

---

[4] For reasons which will be concerns of the present essay, Newton's
work on chronology, some of which was published posthumously in
his *Chronology of the Ancient Kingdoms Amended*, is intimately linked
with his theological studies, including his published *Observations upon
the Prophecies of Daniel and the Apocalypse of St. John*, which appears in
Horsley's edition of Newton's works: Samual Horsley, *Isaaci Newtoni
Opera Omnia* (vol. 5, 1785). A small but most interesting collection
has long been available: Herbert McLachlan, ed., *Sir Isaac Newton:
Theological Manuscripts*. See the extensive discussions in the Westfall
biography, and Castillejo, op. cit., chaps. 2, 3, and 4. Frank E. Manuel
has published studies of both aspects of Newton's thought: *The
Religion of Isaac Newton*, and *Isaac Newton, Historian*.

Of what philosophy, then, is the *Principia* the "principles"? What does "natural philosophy" include in Newton's mind? This is Christian natural philosophy, equivalent neither to the philosophy of the ancients nor to the pragmatic enterprise we call "science" in our time. "Nature" is for Newton, I suspect, the field of God's working among us; the *Principia* is a tormented, mystic work, shaped to open the way to this vision. Let us experiment in any case with reading it from this point of view, taking in the present essay what can be no more than a very brief walking tour through certain halls and chambers of an immense and in some way magic work.

The Newtonian Alchemy

Let us begin by attempting to characterize certain features of the Newtonian alchemy which will serve as guides to this reading of the *Principia*. It is not possible here to embark on an account of the alchemy itself, a field far too complex and still, for most of us, too obscure to summarize readily. Newton was in some way, we may say, embarked on a systematic and disciplined and, we must add, perfectly rational search for real principles hidden behind the appearances of ordinary materials. For example, the "opening" of common mercury would yield a real, or philosophical, mercury. Once achieved, mastery of such true substances would make possible not only a new order of knowledge of nature but the ability to command practical outcomes which indeed included transmutations. Newton believed that Robert Boyle had accomplished the transmutation of water into earth.

Here, we can only list certain of the attributes of alchemy, a list which will be deliberately selective—chosen with an eye to those features which most tend to shape Newton's understanding of the task of what we now call mathematical physics. For the *Principia* is conceived as instrumental, as mathematical servant to a greater project. And that greater project, natural

philosophy, evidently takes its own shape in Newton's searching mind in the context of the vast work which he was deriving
from the alchemic tradition. As we have suggested, the result is
something which is on the face of it very far from what we have
been taught to think of as "scientific method"—yet perhaps
not so far from the project in fact under way in the laboratories
of the modern world when considered from this alternative
point of view. About this last question we must reserve judgment until we have come to terms with the *Principia* in this new
context.

What "Science" is

It is regularly said that the Scientific Revolution freed men's
minds from shackles of an encrusted tradition, characterized
as the weary reiteration of principles drawn from the philosophy and logic of Aristotle. In the figure of Simplicio in Galileo's
Dialogues, we see well enough the kind of complacent common
sense to which this reproach applies. Flying under other flags,
the type is familiar in any age. But we should be careful not
to underestimate the influence, and possible truth, of principles articulated by Aristotle and conveyed in Newton's time
by better men than Simplicio. However true it is that Newton
as a student at Cambridge turned to the new "mechanical philosophy" and left blank the pages in his personal notebooks
reserved for the traditional topics, certain understandings
concerning the role of mind and the nature of the sciences
may have been more deeply implanted. What we turn from in
our youths we tend not to leave altogether behind but rather
incorporate in new structures. Such is the dialectic of learning,
and Newton at Cambridge was engaging in intense dialectical
maneuvers. Alchemy in certain fundamental features borrows
concepts of Aristotelian science, and so, I would suggest, does
the *Principia.*

Aristotle defines the term *science* in that culminating section of his logic known to tradition as the *Posterior Analytics*.[5]

The essential point is that "science" concerns that which we know best: knowledge which is scientific is certain, and not hypothetical. The exemplar is our knowledge of mathematics—arithmetic or geometry. The sum of two and two is four: that is not a contingent or hypothetical statement. The principles on which a science such as geometry is founded are correspondingly evident: to see them is to grant them. They are of the sort Descartes calls "clear and distinct." But in human practice, we have to learn even these mathematical sciences through experience, first encountering individual instances, and only gradually coming to perceive in them the universal principles which they illustrate. That coming-to-see of principles which are ultimately evident is what Aristotle calls "induction." There is nothing probabilistic or insecure about such induction, and its culmination is not in a statement which is tentative or (in the terms of modern technicians of these matters) "corrigible." Therefore it is not much like what is called "induction" in our talk about the experimental sciences, in which induction is understood to yield principles which are more-or-less likely, and which stand only until they are tested and corrected or revised by experiment. They are never certain, and even the most widely supported of them are ultimately contingent and subject to the next major revision of our scientific paradigms. We have seen too many revolutions in the foundations of even the most secure sciences to take any current offering as final; to think otherwise is seen as dogmatism, closure of the mind to the prospect of new and unexpected empirical evidence. Whether we are simply right in this, or whether Aristotle has something yet to teach us, we need not resolve at this point. It is enough to note that Newton and the alchemic tradition

---

[5] Aristotle discusses induction and explains that scientific induction leads to the kind of knowledge which is "always true," at the close of the *Posterior Analytics* (100b 5–10).

stand in this respect on the side of the ancients: for both, disciplined induction can lead to absolute truth.

Alchemy fundamentally extends the Aristotelian assumption to include nature as a whole; in principle, the whole body of nature becomes knowable as only the mathematical sciences had once been. Underlying the shifting phenomena are constant factors, principles which in their hidden operations generate the world as we perceive it. These principles, obscured as they are beneath the veil of appearances, can become objects of knowledge. Fetching the true and philosophical reality out of the confusion of common appearances requires high art, art which in the traditions of alchemy, for reasons not necessarily misguided, was passed on in secrecy and mystery. Such knowledge, the great goal of alchemy, would be not contingent but as certain as any object of an Aristotelian science.

Before going further, it is important to distinguish more carefully: Aristotle was no alchemist! Alchemy, as an alternative tradition, extended to all things the certainty which in Aristotle belongs only to the objects of intellect. At the root of the distinction between alchemy and Aristotelian science lies the obscure concept of matter, or more generally, the character of the material world. For Aristotle, matter is a shifting and in itself unknowable principle, which in its mingling makes of the world a problematic realm. Matter as the "principle of individuation" brings the universal to bear at a place or at a time, and with such individuation comes an inherent obscurity. Scientific knowledge therefore is not of individuals but of the forms which give shape. Alchemy proposes by contrast that hidden beneath this flux are knowable principles embodied in real and philosophical entities. Material substances, characterized in such terms as the *philosophical mercury* or the *philosophical sulfur* are real entities which would be knowable if they could be fetched out. To know them would in turn be to gain power over phenomena, to secure at once knowledge and command of the realm of nature. As we shall see in more detail shortly,

Newton thus speaks as an alchemist throughout the *Principia,*
as he frames his arguments and his experiments to find out the
true and philosophical behind the complications of common
appearances.

Evidently, the watershed which divides alchemy from the
philosophy of either Plato or Aristotle is the Judeo-Christian
God, and the account of Creation. That "matter" which for
Aristotle blurs being in time or place is no blur when it is
made by God: with the idea of Creation, we may say, the fact
takes primacy over the universal. And that is a leading thread
we may borrow from alchemy in our reading of the *Principia*:
for Newton as well, fact has primacy over universality. In this, a
theorem of the *Principia* is not at all like a theorem in Euclid.
A proposition of the *Elements* is a truth, to be known, contem-
plated, and prized as such. A comparable proposition in the
Principia is instrumental. It is universally true, but it is interesting primarily as a means of finding out truth of another
kind, truth which exists in this world as philosophical and
knowable Being. The *Principia* is essentially a Judeo-Christian
work. The truth, the "philosophy," to which its theorems are
instrumental is the truth of God's concrete work in the world.
We see, then, that here alchemy, theology, and history belong
comfortably together; they are facets of what Newton calls
"philosophy."

We are on the track, I believe, of the Newtonian alchemy.
"Natural philosophy" thus understood becomes science—not
in the sense of our modern convention, according to which
"science" is contingent theory entrained to ever-evolving
empirical evidence, nor simply, either, in the sense of the
Posterior Analytics, in which "science" means knowledge as universal—but in a third and new sense, close in its certainty to
Aristotle, but giving primacy to truth as fact in the spirit of a
created cosmos. We know now that Newton pored over the
alchemic tradition, mastering enormous bodies of esoteric
literature, comparing and cataloging terms, revising treatises,
imbibing the powerful symbolism Carl Jung would later come

to recognize among the archetypes. As he did so, Newton felt himself to be on the track of a science in this new sense. And it is such science—science as certain as classical geometry and arithmetic, but now directed to objects incarnate in matter—for which the *Principia* is designed to serve as instrument. The term *incarnate* is appropriate in this context, no less so for its theological overtones; for indeed, with the entrance of God's truth into the world of flesh, Christ becomes the paradigm. Alchemic symbolism is appropriately suggestive of Christian iconography.

Truth in the classical sense is called *mathesis*: strictly formal and teachable knowledge. If we are to find such truth in the alchemist's laboratory, then it will not be out of order to envision "mathematical principles" of alchemy, and indeed we see that that is just what has happened with the *Principia*. Newton has given alchemy its mathematical principles, in good theoretical form. What we shall try to show is that in the *Principia* the mathematics is ordered and shaped to point back to the Creation, the universal thus servant to the fact of God's work in the world.

The Rhetoric of Interpretation

We today are imbued with the spirit of "progress," however our sense of this may have been lately eroded by concerns about global warming and the problem of closure of our proliferation of technological waste and pollution. We have looked to Newton, and the Scientific Revolution so closely associated in our minds with him, as the very models of the possibility of progress in human history. But the sense of these matters is very different in alchemy, and I believe Newton understood the *Principia* in the alchemic sense. On the whole, the long Western tradition has thought of the course of history rather as decline than as progress. Much of the excitement of Renaissance thought, as the term implies, concerned the recovery of something which had been lost: there had been a better time, when people knew more. The Renaissance was encouraged in this by passages in the Greek texts it was finding, which themselves looked back mythically to a Golden Age. Alchemy tends to make this assumption in terms of the Hermetic tradition of an earlier learning, a *prisca scientia,* and Newton accepted this view, not without complication, but ultimately perhaps without real question. It was perfectly plausible for him to accept the conviction of the alchemists that they were bearers of precious texts, the meaning of whose symbols had to be retrieved, embodying wisdom once known but now lost and recoverable only through intense and resourceful study of documents which had come down from ancient times.

Newton works with a sometimes feverish sense of discovery, and it may be that things he came to know he believed had never been known in the form in which he was articulating them—yet all of this is in the context of a sense of human time which relates back to earlier wisdom. It is not a ridiculous point of view, certainly, for the paradigm is the Scriptures themselves. Many of the most respected minds of all Judeo-Christian epochs have devoted themselves to the interpretation of Scripture, with the conviction that no higher learning could be achieved than the fullest realization of the wisdom contained there.

As has now been fully demonstrated, Newton spent a very large portion of his most rational energies in just this work: long, exacting labors at the alchemic furnace, matched by imaginative, almost furious efforts to extract the meaning of a very large body of alchemic texts—some published, many circulated only in manuscript form; some ancient, others still being produced in his own time. Since such *interpretation* is part of the work of classical rhetoric, what this means, especially for our reading of the *Principia,* is that Newton's work belongs to a branch of the art of rhetoric—and again, we are reminded of Aristotle and the university tradition.

Classical rhetoric distinguishes two branches of its art, the composition of new works and the interpretation of the works of others. For the classical rhetorician, it is clear that the art of composition held the highest place—the Orator was the exemplar of the man of virtue in society. But with the advent of the Judeo-Christian faith, sacred text took priority, and the bias within the art of rhetoric shifted from composition to interpretation. Augustine in *De Doctrina Christiana* (*On Christian Doctrine*) sets the new mode, and Newton's work is to be understood as falling within this aspect of the rhetorical tradition.[6]

A very great part of Newton's labors thus went into the highly artful interpretation of texts. As we shall see, he elaborated his own new interpretations of certain Judeo-Christian texts, and he labored endlessly over the interpretation of the texts of alchemy. In the *Principia* two great classical streams thus flow together: it is a mathematical work, but as interpretive, it employs mathematics in the service of the art of interpretation—thus, mathematics in the service of rhetoric. We shall see as we examine the methods of the *Principia* just what this mathematical rhetoric will look like.

The texts in question—both those of the sacred tradition, and those of alchemy—involve a special task not met by the classic rhetoricians, and again, we may take Augustine as guide to the new challenges involved. For classic rhetoric was addressed to the expression of objects to which the human mind was inherently adapted, while both the sacred and the alchemic texts addressed objects intrinsically hidden. Words, signs, and metaphors in Aristotle have major roles, but these are functions which are somehow bearable. Even Aeschylus, speaking the unspeakable, finds metaphors which contrive to bear their burdens. But no metaphor is adequate to the

---

[6] Augustine's discussion of the rules for interpretation at the outset of this text may be compared with Newton's, referred to in the next note.

God of the Jews and the Christians; God does not admit comparison with other objects of thought. In place of metaphor, which as the Greek word it is derived from suggests "bears thought across" by comparing object with like object, we meet symbol which must be understood in another, mystic, sense. Interpretation cannot in this case take the direct path but becomes in itself the art of elaboration of interpretive structures, the penetration of mystery to break through to hidden meanings.

Thus Newton was comfortable operating with the mythic symbolism of the alchemists, and he did not think it unscientific to meet the alchemic authors on their own grounds. The symbols of the lion, the dragon, allusions to the constellations, Leo and the star Regulus, were not out of order where the objects were of the sort Newton understood alchemy to address—hidden powers, active spirits, male and female, principles of growth and decay. For alchemy, the paradigms are first of all organic, the symbols are borrowed from life processes, and this again I think we can assume did not strike Newton as unscientific. As we shall see, these organic and psychological processes are just the matters he was concerned to bring within the scope of his physics. Freud and Jung are close in principle to the heart of his project: Newton's regular symbols are, as we have suggested, among Jung's archetypes, while the energies he is concerned to manage—the prototypes of the Newtonian "forces" of the *Principia*—are close kin to the energies Freud recognizes as surging in the psyche. We tend to think of the *Principia* as laying the foundation of *physics* in the modern sense of the term, but if we are correct in the present claim, the customary view of Newton's work is far too limited. Rather, Newton is most concerned with questions of biology and the psyche, with vital principles at the foundation of organic process and the functions of the psyche. Application to the motions of the planets was only the first and most available application of his principles; all of nature and the mind of man lay ahead.

The Mathematical Rhetoric of the *Principia*

Let us turn now to Newton's text, beginning with his Preface to the Reader. On the whole, Newton writes with a grammatical and rhetorical precision which rewards careful reading. We might note that the standards, or rules, for his own writing may be understood as the inverse of the rules he formulates for the interpretation of a text: everywhere the arts of language and its uses—the arts which tradition calls those of the trivium—have these two faces. As arts of composition, of the making of discourse, they are guides to speaking or writing; while as arts of interpretation, they are guides to listening or reading. For Plato and Aristotle, as we have suggested, emphasis was on the arts of making; with Augustine and the Judeo-Christian tradition, a transformation takes place by which, in the presence of sacred text, the art of interpretation necessarily takes first place and becomes greatly and subtly elaborated. Newton is working in this latter tradition. For him, human learning is a question of the interpretation of immense bodies of text. These are first of all two, sacred Scripture and the phenomena of nature; the Bible, and the Book of Nature. Beyond them lie, as we have seen, the texts of deep traditions from earlier learnings, original *prisca theologia* and *prisca scientia,* gathered as the classics of theological interpretation and, of special interest to us here, of alchemy.

Readings of Scripture take us first to moral law and theological truth, while readings of the phenomena of nature take us to laws of nature and the understandings we call philosophy—but the two reflect fundamentally and intricately on each other. What we may call the Newtonian project is shaped to the conviction that Truth is one. Thus it is not surprising that when Newton formulates rules of procedure for the new adventure of mind on which he is embarking, the rules for the interpretation of Scripture, the *regulae interpretandi,* and those for the interpretation of the Book of Nature, the *regulae philosophandi,* mirror one another. Among the vast number of documents which Newton left unfinished or unpublished are two which set forth these new guides for the working of reason in the two faces of its interpretive mode.[7]

At the beginning of one work, on the interpretation of prophecy, he set out what were denoted explicitly "rules of interpretation"; and at the outset of the third book of the *Principia,* called "The System of the World," he set out the other set of rules which, though they are called "rules of doing philosophy," are really rules of interpretation as well.

Our mathematical physics is thus legitimate heir to the long and rich tradition of the art of rhetoric, in its aspect as the art of interpretation. Francis Bacon had seen this clearly—the crucial experiment, the *experimentum crucis,* lies at the cross-roads between alternative interpretive paths, highways of thought we now call "paradigms" of scientific theory. These rules of interpretation interweave as well with rules of composition: the *Principia* is at once a careful and exact composition on Newton's part, and the outcome of a new interpretation of nature. It is a composition which formulates a new method of interpretation, using mathematics to unveil realities not before seen. In this sense Newton's intention is to forge a new kind of instrument to serve the needs of a new understanding of philosophy, a philosophy in accord with what we are calling "the Newtonian alchemy." He designs the *Principia* to articulate this new way and thus to effect a turning, a dialectical reversal in the liberal arts. The *Principia,* as we are seeing, looks both ways: in its propositions the arts of the past and the arts of an envisioned future are suspended in extreme tension. We now call this state of intellectual and social tension the "Scientific Revolution." Perhaps a reading of Newton's subtle text in this rhetorical aspect helps, as we have suggested, to throw renewed

---

[7] See Newton's discussion of "The Language of the Prophets," in McLachlan, *Theological Manuscripts,* pp. 119ff.

light on that complex revolution, to which we are the uncertain heirs.

Newton's Preface to the Reader

In the Preface to the Reader, which he wrote for the first edition of the *Principia,* Newton gives a very tight account of the revolution he intends to launch upon the world. The first sentence already carefully, if very schematically, locates the *Principia* in the history of natural philosophy. Newton first characterizes the ancient investigation of nature as culminating in a certain understanding of "mechanics"—he means most of all the mathematics of Archimedes. He then alludes in a phrase to a long middle period in which the trivium had displaced the quadrivium, the arts of language displacing those of mathematics. Mathematics, was set aside and other terms of analysis were substituted, "substantial forms," he says, and "occult qualities." The third, modern period, he sees as a restoration of the old—literally, a recalling of natural philosophy to its mathematical mode, yet in a new guise of what he calls mathematical "laws." We should take note of that term, for it bespeaks the passage we are embarking on. Mathematics in its first instantiation had spoken by way of theorems, universal statements addressed to the contemplative mind; now its utterances will be laws addressed not in the first instance to mind but in the form of commands to action, directed to an obedient nature. This history—the whole intellectual history of Western man in relation to the study of nature—has been set out in a single dependent clause: we may think of it as the "whereas" clause of Newton's own proclamation. Let us try restating it in the form of a manifesto:

> WHEREAS the ancients … most of all prized mechanics in the investigation of nature … and those of more recent times, having discarded substantial forms and

occult properties, have undertaken to recall the phenomena of nature to mathematical laws, ...
IT IS PROPOSED in this treatise to cultivate *mathesis*
insofar as it looks to philosophy. (Cf. *Principia*, p. xvii.)

This is the very form of a declaration, a manifesto launching
a new intellectual endeavor. Others, Newton says, have gone
far: but they have failed to achieve the full union of *mathesis*
and philosophy, and it is this which the *Principia* will teach
mankind to do. The first word of the Latin title is *philosophy*.

We should notice that his three-part account of history is in
the precise form of the dialectical process, as we meet it first
in Socrates, but ultimately in Hegel and Marx. There is first
the affirmation in the mode of innocence—here, the initial,
brilliant thrusts of the ancients, Euclid and Archimedes. These
gave us universal propositions, addressed to timeless mind.
Then there is the medieval negation in the name of spiritual
aims and an omnipotent, omniscient God, mysteries which dispelled such ambitions of the merely natural intellect. Now at
last there is to be Newton's synthesis in a new mode. As we
shall see, this will empower a new *mathesis* and the old intellect but will incorporate spirit and mystery, to go beyond mere
universal mind to address all things in a nature recognized as a
Creation infused with the operations of divine agency.

Newton uses the Greek term *mathesis* in his declaration,
to suggest the scope of the enterprise: he means demonstra-
tive argument and the cultivation of intellectual intuition in
a new sense, and in a new application. Though he embraces
Euclid, Apollonius, and Archimedes as teachers, we shall
see that he boldly extends the bounds of their geometry to
become something he will call "universal mathematics"; and
though his argument in the *Principia* will on the whole be in
the form of demonstration, his intention is to transform altogether the concept of a mathematical treatise. Rather than
a theoretical composition such as Euclid's *Elements*, setting
forth what is known and unknown for the ultimate benefit of
an understanding mind, learning as an end in itself, Newton's

work is designed as instrumental. Offering its propositions as instruments of an interpretive inquiry of an altogether new kind, it is to transform the nature of both mathematics and philosophical investigation. Its *mathesis* is to open anew all studies, studies directed to a Creation which is throughout, *mathesis* incarnate.

The "opening question" of the *Principia*, then, must be this puzzle: "How can *mathesis* bear upon (or 'look to') philosophy?" In contemporary terms, we might tentatively put it this way: "How can a broad philosophy of the Creation find its principles in, or through, mathematical physics?" In terms of our present intellectual scene, we may wonder, "How can the mathematical sciences and their corollary, a technological society, be brought back into the circle of philosophy—of human purpose, and the understanding of ultimate things?" Common wisdom today assumes that it can't be done—that our science or *mathesis* is of one sort of thing, which we may in some sense come to know, while the larger circle of philosophy and theology escapes *mathesis*, and hence deals with things about which we may have opinions or beliefs but cannot resolve as matters of knowledge. We see how far we may be, or suppose ourselves to be, from the endowment Newton meant for us. We seem in some way to have missed Newton's intention.

Having in his opening declaration given this epitome of the problem he has set for himself and the world, Newton now takes some time to consider what the extended and inclusive *mathesis* of the *Principia* will be like. To achieve it, Newton must first bring about a revolution within mathematics itself; thinking in terms of the traditional seven liberal arts (the four mathematical arts of the quadrivium, together with the three linguistic arts of the trivium), we might say that Newton, in order to bring the quadrivium into an effective new relation with the trivium—to get *mathesis* actually to serve philosophy—must first perform a revolution within the quadrivium itself. The pivotal term for Newton at this point is *mechanics*. The

ancients, even when they brought the full power of their
mathematics to bear upon what they thought of as mechanics,
considered it under the paradigm which ultimately Homer had
given them, and us, when he called Odysseus *polymechanos*—a
general-purpose mechanic, or a man of many "devices." In this
sense, Newton says mechanics was for the ancients a realm of
human forces. When Archimedes, as the ultimate mathematical Odysseus, proposes to use the lever to move the world, the
lever is a machine—one of the five classic "machines" (the lever,
the screw, the inclined plane, the wheel, the pulley)—but the
force upon it is the beast of labor or the hand of man. Newton
sets up his own view in dialectical opposition to Archimedes in
this way:

> This part of mechanics was cultivated by the ancients,
> with an eye to manual arts, in terms of the five powers,
> and they considered gravity—since it is not a manual
> power—hardly otherwise than in respect to moving
> heavy bodies by those powers.
>   We, however—consulting not arts but philosophy,
> and hence writing not of manual but natural powers—
> treat most of all of those which look to gravity, levity,
> elastic force, the resistance of fluids, and forces of this
> kind. (Cf. *Principia*, p. xvii.)

Mechanics, then, had been seen as belonging to human
arts and the realm of human *praxis,* the lower realms, we might
say, of that divided line according to which Plato organizes
the hierarchy of human functions in *The Republic*. Newton
proposes to transform mechanics to much higher philosophical status, by recognizing in nature—that is, in a domain
which is not of man's contriving—"forces" or powers which
we can understand by a sweeping analogy to those human arts
and powers of Odysseus and Archimedes. Here for Newton
nature imitates art, in the sense that we move from the classical mechanics of human machines to a new, philosophical
mechanics of natural powers, in which nature—or some

agent which moves in and through nature—is the artificer.
What were human arts are to be seen now as divine workmanship: God is the new artificer.

We see, then, the prospect of a philosophical mechanics. To
relate this to the classical quadrivium of arithmetic, geometry,
music, and astronomy, however, we must back up to look at an
earlier section of Newton's Preface. Newton has made the following claim:

> Geometry is founded therefore in mechanical praxis,
> and it is nothing other than that part of universal
> mechanics which exactly proposes and demonstrates
> the art of measuring. (Cf. *Principia*, p. xvii.)

Here Newton is joining issue with Euclid and arguing that
what Euclid takes for granted—demands, or "postulates," in
the language of the *Elements*—or asks the beginner in his art
to grant, in fact derives from another science: namely, the
de-scription, the scribing-out of the circle and the straight line.
For Newton, this small distinction is not a quibble but a fundamental clue. These two operations upon which geometry must
be built he here calls *praxis,* again perhaps using the Greek
term in order to open a concept anew. *Praxis,* from the Greek
verb *prattein,* means a "making" or "doing," but this geometric
scribing is not the kind of doing which belongs to the artifi-
cers at the bottom of Plato's line. It must be a making which
is higher in the intellectual hierarchy than geometry itself—
that is, it must be as clear to the mind as the propositions of
Euclid themselves, and prior in the order of understanding. If
it were not, geometry would be founded on a clouded insight.
Newton is thus proposing that there is a *praxis* which is a higher
mathesis, and that by confusing this with ordinary mechanics
the ancients overlooked a fundamental possibility. He says:

> However, the errors are not of the art, but of the
> artisan. Whoever works less accurately, is a less per-
> fect mechanic, and if anyone were able to work most

accurately, he would he the most perfect mechanic of all *(mechanicus omnium perfectissimus).* (Cf. *Principia,* p. xvii.)

Clearly, Newton has a candidate for this post of *mechanicus omnium perfectissimus,* who will bridge the erstwhile abyss between the accuracy of even the best human workmanship and the exactness and intellectual clarity of mathematical demonstration. It is God Himself, God the Creator whose word has entered upon the human scene between the time of Archimedes and the time of Newton. This concept of the omnipotent Creator as divine craftsman makes it possible for Newton to pass from imperfect mechanics to a *praxis* so exact that it becomes *mathesis.* The God of Genesis transforms the quadrivium and clears mechanics of its confusion. We must recognize, above geometry in the hierarchy of the liberal arts, a new, dominant member of the former quadrivium: universal mechanics. Within this new structuring of the quadrivium, motion is primary; the mechanics of powers or forces will be one branch, while pure geometry, as art of measurement devoid of the question of force, will be another. We might say now that universal mechanics is the "real geometry," of which Euclid's is one limited aspect. The other part Newton goes on to term "rational mechanics," and this is the part concerning powers, or forces,

> In which sense rational mechanics will be the science of motions which result from any forces whatever, and of the forces which are required for any motion whatever, exactly enunciated and demonstrated. (Cf. *Principia,* p. xvii.)

Here, in this brief definition, we see a link forged between the new *mathesis* and philosophy. Essentially, I believe, Newton agrees with Aristotle in one aspect of his understanding of natural philosophy: namely, that the object of natural philosophy is motion—those motions which occur in the natural order, regularly and without man's assistance. The object of the natural philosopher is to discover the true causes of these motions. In implementing the new program for the conversion of natural philosophy, the initial difficulty must be to pin down a concept of motion at the level of strict mathematics; the next will be to come to an agreement about "cause."

It would seem, as indeed it seemed to the ancients, that the concept of motion inherently resists any effort to achieve intellectual clarity. If an object of thought is to be clear, it must first come to rest before the mind—but motion is exactly that which never rests. *Mathesis* would seem essentially addressed to *stasis*; *kinesis* would seem to be by its nature to defy strict knowing. Aristotle led the ancients to a solution of this problem in his *Physics* by means of a brilliant grammatical artifice. He captures the concept of motion—not mathematically, it is true, but adequately for a philosophy of nature committed to operating in a less mathematical cosmos—by means of a double predication of being: to be potentially, and to be in actuality. The motion of the growing tree is caught by seeing double: once *in potentia,* as acorn, once again in actuality, as the mature oak. This is no small triumph of the ancients, this old way, for it has the power of showing us the motion in its wholeness, and it applies across the board—to falling bodies, or in the *Poetics,* to falling kings. It was the old way, to grasp the concept of motion through the instrumentality of the trivium. Ptolemy, you might object, did more through the concept of uniform motion, the ratios of elapsed times and interpolation in tables—but always he is dealing with the relations between finished motions, not with the planet in the process of the motion itself; he tells a story but does not claim a cause. Newton is intent on going directly to motion itself, not such results of elapsed intervals, as the primary object. How he achieves this we shall see shortly, when we turn to what he calls his "lemmas," but first we must consider Newton's treatment of the question of "cause."

In the domain of cause, Aristotle's *Physics* distinguishes four modes of answer to our question, "Why?"[8]

The response may come in terms of the matter, the agent, the form or definition of the thing itself, or the end or goal of the motion. Aristotle's Greek is always simpler than our English: these are, respectively, the *in-what*, the *by-what*, the *what*, and the *that-for-the-sake-of-which*. Again, Aristotle is using the instruments of the trivium, projecting nature into language by means of crucial and indeed highly effective distinctions of grammatical structure. And again, in a world which is the workshop of a different God, Newton carries the question into the realm of the quadrivium, in this case by locating the issue in a single concept of cause, in Latin *vis*, meaning "power" or "force," and claiming that the relation of motion to its cause— of motion to force—is a problem for solution by *mathesis*, exact enunciation and demonstration. For Aristotle, the relation of cause to effect in nature (and nature is the source of all action for Aristotle, organic, inorganic, vital, or mental) is orderly, but blurry, for matter, that principle of confusion, dispels precision. For Newton, nature in effect snaps to attention, precisely because behind all action is that divine agency which imposed law upon its Creation.

While we are still speaking of Newton's purposes, and before we turn to look at the new mathematics which will be the key to his method, we might well pause to look ahead to the actual structure of the *Principia*. The *Principia* is organized in three books. The first two contain the body of basic propositions which constitute what Newton calls the "mathematical principles of natural philosophy"—strictly, perhaps, these two books are the *principia*, the "principles." The third book, called "The System of the World," is then appended to the work as

---

[8] Aristotle discusses the possible answers to the question "Why?" in Book II of the *Physics* (198a 15–24). Newton's *Principia* may perhaps best be thought of as founding a physics which is ultimately coextensive with, and an answer to, Aristotle's (i.e., closer to biology and psychology than to what we call "physics"). The world-change between the two may be thought of as the conversion of the question "Why?"

an illustration, or one might say, an initial realization. Where the first two books, as the books of the new *mathesis*, only await their application in philosophy, the third book is a brilliant example of that new philosophy itself. The first two books are thus books of mathematics in the new mode; the third book is an exemplary book of philosophy. As such, the third book is the first step on a way which is ultimately, I think, intended by Newton to yield a total replacement for Aristotle's *Physics* in its full range. Newton aims to be forging here a *mathesis* adequate to the one truth of a created world.

The sample in the third book deals with only one of those natural forces Newton has alluded to, namely gravity, and addresses the cosmos only in its aspect as a gravitational system. It conspicuously does not, for example, deal with the optical or chemical systems of the world, nor with the causes of the vital motions of nature, those which alchemy calls "vegetable" and "animal" and is primarily interested in mastering. We recall that "physics" no more denotes to Newton than it does to Aristotle that "inorganic" realm which we now call physics and distinguish from biology and psychology. Newton clearly intends a mathematical physics of all natural things, including most especially all those which live, grow, move, and feel. The gravitational system unfolded in the third book is a brilliant but relatively easy initial step into the new philosophy. The cosmos is only the beginning. The real third book, which it was not possible for Newton to compose, would be the Book of Alchemy. There is evidence that Newton was trying hard to bring his alchemy into shape in time for inclusion in a more complete picture of The System of the World, balancing the macrocosm of the planets with the microcosm of a mathematical alchemy. Newton was not able to write the *Principia* he intended. One clue to his ambitions is found in the great "Queries" appended to his *Opticks*, in which he did his best to pass his vision on to future generations. It is a vision of alchemy realized.

We have seen that Newton proposes to bring the investigation of physics in this broadest sense into intimate relation with a transformed version of the quadrivium, the arts of mathematics. We have seen the aim, but not a method by which it might be brought about. The key term will be *force*, for Newton is shaping

> ...the science of the motions which result from any forces... and of the forces which are required for any motion.... (Cf. *Principia*, p. xvii.)

If "forces"—we will reflect later on the possible meanings of this most mysterious term—are causes, we see that the science of mechanics will proceed by two great modes of argument. In one, we proceed from motions to find out the forces—from effect to cause; the other will be the reverse, passing from forces to resulting states—from cause to effect. The first moves in the direction of induction—from the phenomena to find out the underlying principle. The second moves in the direction of construction from the principles well known, to demonstrate the effects. In the rhetorical terms we have employed earlier, the first corresponds to the rhetoric of interpretation—interpreting the appearances to find out what lies beneath; while the second belongs to the rhetoric of composition. Both are now to be conducted by means of mathematical demonstration. Newton goes on to formulate this plan in a larger way:

> Every difficulty of philosophy is seen to turn upon this: as from the phenomena of motion we investigate the forces of nature, so thereafter from these forces we demonstrate the remaining phenomena. (Cf. *Principia*, pp. xvii–xviii.)

When we see that the "motions" in question are to include all the phenomena with which nature presents us, and that the passage from the motions to the forces is the intellectual progress from effects to their causes, we recognize in other terms that Newton is referring here to that progression from result to cause which Aristotle calls *analysis*—that is, the direction of inquiry in physics. The reversed process, the passage from the causes, now known, to their consequences, is the direct motion of philosophical argument, the building of a system of consequences from a few first principles which are themselves best known. This is *synthesis*. Euclid's text is a work of nearly pure synthesis, and in the tradition of the *Elements,* we normally think that progression, from elements to outcomes, is the appropriate mode for demonstration. But what Newton is proposing is that both of these two great motions in philosophy—the order of inquiry and the order of knowing—be carried out by demonstration in the new *mathesis*. He is not however the first pioneer to cross this dramatic watershed, for he is very much the student of Descartes.

There are as well certain occasions when the ancient geometers proceed the other way: they start with the unknown and reason toward the known. How can that be done? The formula is, in effect, to "assume the thing done"—that is, take the unknown as if known. In the case, for example, of a geometric construction, suppose the figure which is sought actually to have been constructed; then, on this assumption, reason to those necessary consequences which would follow—reason in the subjunctive mode—until you arrive at some consequence which you know in fact is true. From this basis as *terra firma*, if the steps of the argument will admit of reversal, the demonstration can be stood on its head and, run backward, turned into a synthesis. At the end of it all, the unknown is arrived at again, now as a necessary consequence of what is indeed known to be true.

All this is familiar to the ancients, though on the whole they use strict analysis very seldom. Since that was widely believed to be the method by which they had hit upon many of their synthetic propositions, it was thought of as the method of discovery, and scholarly rumor held over the ages that the ancients had been in possession of such a systematic art of discovery which they had kept secret, disdaining to pass it on to posterity. It was perhaps the principal mathematical component of the tradition of a lost *prisca scientia*. Descartes speaks of it in these terms and proposes at the outset of his *Discourse on Method* to recover this presumed ancient art of discovery though in a new form, more powerful than anything the ancients had possessed: a universal method of reasoning from the unknown to the known.[9]

In the broadest terms, his *Discourse on Method* sets forth the new universal analysis in the guise of the method of "doubt." In its mathematical implementation in relation to problems from Euclid and Apollonius, he calls it analytic geometry and appends it, as an illustration, to the *Discourse*.

Now the idea of a systematic mathematical analysis, by which to move by precise demonstration from unknown to known, is the very methodological backbone of the *Principia*. In this, Newton is altogether a disciple of Descartes. Yet what Newton gives us is not the expected algebraic analysis but a *geometric* analysis of his own devising, and this represents from the outset a significant, and very deliberate, turning away from Descartes. Newton is a master of Descartes's algebra and uses it whenever he must; he knows well what vast new powers it opens in mathematics. But he knows too that those symbols, letters representing the unknown or representing nothing, mere symbols, reduce science to an obscurity and move by way of automatic processes which bring no light to the mind. It is for this reason that there is a minimum of algebra in the *Principia,* and in its place a maximum of appeal to the eye of the mind through the artful use of what we might call a rhetoric of geometric forms. This is a true battle of the arts: from Newton's point of view,

---

[9] Descartes discusses the analytic method of the ancient mathematicians, and its relation to his own method, in the *Discourse on Method*. He appended his *Analytic Geometry* as an illustration—perhaps much as Newton appends the astronomical "System of the World" to the *Principia,* as just one particular application. It is important to note that the method of algebra is the paradigm for the sweeping method of "doubt" of the *Discourse* and the *Meditations*.

a battle against the new, illiberal art of algebra, and an effort to reinstate the geometry of the ancients, though now as the vehicle of a powerful and precise analysis as well as a synthesis.

We shall see that the *Principia* is in one most important aspect a work of polemic, against a view of the world which Newton regards as deeply pernicious. The enemy, to whose refutation Book II is chiefly dedicated, is at once Descartes and that view of the world, based on Descartes's work, known and widely prized as the "mechanical philosophy." Given our modern carelessness in the use of terms, and the fact as well that Newton is founding the *Principia* on a mathematics which he is calling a "universal mechanics," it may seem strange to paint him now as the enemy of the mechanical philosophy. The crux of the mechanical philosophy is that it reduces the workings of nature to those of a machine in the sense of a clockwork, connected throughout by linkages and gears, and actuated always by parts in direct contact. Now, algebra is such a machine in the domain of the mind. The processes of algebra turn like the gears of a clock and thus proceed without dependence on or reference to any meaning of the symbols involved, which are in their algebraic role mere blind counters. Algebra is, then, the appropriate projection of the mechanical philosophy into mathematics: and as such, Newton excludes it from the *Principia,* not as an incidental tactical decision, but as a deep rhetorical principle. The *Principia* is aimed not at problem solving, however our modern world may have adopted it in only that sense. It is meant as a philosophical instrument, and as such, it must be an instrument composed in symbols which illuminate and clarify, as the artfully devised figures of Newton's geometric analysis will in fact do. This is for Newton a decision of deep interest, and it already points toward the largest significance of the *Principia.* It is directed in its substance against the mechanical philosophy, and in its form it defies the algebra of the *Geometrie Analytique.* These are decisions made on theological grounds, for the mechanical philosophy and its mathematical corollary are what Newton in his theological discussions calls "dead works." Turning his *Principia* against them is for Newton a most intensely motivated move in the service of God, and in defiance of Satan. These are to be the mathematical principles of a knowledge of God's works. Both substance and style of the work must therefore be appropriate to the service of God, as the mechanical philosophy and its algebraic mathematics are not.

Newton agrees with Descartes that the world is a mathematical object—an appropriate object for *mathesis*—to be approached by methods of mathematical analysis; to this extent, Newton and Descartes concur. But for Descartes, this means that the domain of nature is a simple object, clear and distinct in the mind's eye. "Matter is extension," Descartes tells us. Descartes means by "extension" precisely that magnitude, *megethos,* which is the subject of Book V of Euclid's *Elements*.[10]

It is a mathematical object of which one can indeed, in some sense, form the kind of "clear and distinct" idea, which is the proclaimed object of the Cartesian science. It might seem, indeed, the perfect candidate for a mathematical interpretation of the concept of "matter," since magnitude is in itself at the same time quantitatively precise, and utterly without

---

[10] Descartes's analysis of the idea of matter proceeds by way of his reflections which doubt away the qualities of a piece of wax (*Meditations*, Meditation II). What is left is "magnitude or extension in length, breadth or depth" (Meditation III). Euclid's general theory of magnitude, which Descartes would have particularly in view, is given in Book V of the *Elements*. Descartes's definition of matter is set out as one of the first principles of material things in his *Principles of Philosophy*:

That the nature of body consists not in weight, nor in hardness, nor colour and so on, but in extension alone.

(*The Philosophical Works of Descartes,* trans. Elizabeth S. Haldane and G. R. T. Ross.) Descartes spells out the world in terms of this plenum of extension in *Le Monde,* the book which is the system of his world: *Le Monde: ou, traité de la lumière,* trans. Michael Mahoney.

form or definition of its own. But for Descartes, this extension in motion constitutes the totality of the natural world. All bodies are configurations of this one matter, all motions are its motions, in the form of flows and vortices of every kind. This is indeed the world of the mechanical philosophy—empty in Newton's view however much it is full, "dead" however much it moves. It invites the arrogance of the calculative mind which can track its every possible complication and freely devise a hypothesis to account for any possible phenomenon. Newton's Christian *mathesis* of the world must therefore be in some way the opposite of Descartes's.

In place of this blind and dead clockwork of Descartes, Newton unfolds a world which is a proper object of *nous*, intellectual intuition, full of intelligible causes and evidences of purpose, plan, and drama. These are objects which are of the highest importance to comprehend. The philosophical vision which Newton's geometric analysis is shaped to achieve shows the mind the way to true causes, which will be at once both material and spiritual. In this sense, the real model for that mathematical analysis which Newton has in mind is the Christian rhetoric of interpretation, by which the Creation is to be read as our evidence of the presence and working of the Creator. Any such reading of the Book of Nature, meant to entail a transforming passage from flesh to spirit, requires art and sensitivity and does not proceed by smooth stages but rather goes through a succession of crises in the construction and interpretation of signs. It is Newton's task in the *Principia* to construct geometric signs matched to the signatures with which the Creator has signed the phenomena. Newton's project is to show that even such an interpretation may be carried out by mathematical analysis when that is couched in geometric hieroglyphs of the right contrivance. The crises of interpretation may take the form of crucial experiments in which nature is put to the test in the laboratory, or crucial observations in which the crisis is transposed to the heavens.

The world becomes a mathematical problem in an interesting sense, whose solution is, ultimately, God.

This is the analysis which the third book of the *Principia* will illustrate in what it is not unreasonable to call its "revelation" of the operation of universal gravitation throughout the frame of nature: every body throughout the cosmos, Newton will demonstrate, attracts every other body with a force of a single genus called "gravity," ruled by an exact, intelligible mathematical law. Without high art and the methods of disciplined experiment, the intellect unaided could never have uncovered such a mystery, though it operates everywhere in our very midst. Its discovery is a dramatic peripety in human history, a breakthrough in man's understanding of the divine plan. As a first step in a new era of natural philosophy, it is a sure sign for Newton that time is running out on the Creation, and man is drawing closer to his God. It has, as Newton's account of the System of the World unfolds, many corollaries concerning our human history, and God's plan for the world.

Time

Between Newton and the classic understanding of reason and the work of mind—between the Lord God, on the one hand, and *ho theos,* the god of Aristotle's *Physics,* on the other—lies the idea of the reality of time. Roughly speaking, we may say that Aristotle denies the existence of time. When he argues that "time is the measure of motion"—and no more—he tells us that substances exist, and they exist most when they are most in act, that is, most in motion. Time is only our count of the cycles of these motions. That is all it is: nothing in itself, only a reckoning of something else, which is. There is no past, there is no future, what is is always present, most present when it is in passage. We live in the midst of memory and expectation, indeed, but memory cannot be of anything essentially different, nor can there be expectation of anything essentially new. For ultimately there is always and only the utter constancy of the unmoved mover, whose necessary existence and invariant act guarantee the integrity of the cosmos from moment to moment and are in the fullest sense the unchanging cause of all its motions. It is a mere corollary of this denial of time that Aristotle denies as well either a creation or a terminus of existence—either a beginning or an end of things. Motion stretches in its potential infinity forever backward, forever forward, but such "forevers" hold nothing in prospect or retrospect. Broadly speaking, Plato, Homer, and the Greek tragedians share this understanding that there are cycles of natural motion and human action, but that time is nothing other than the count of their repetitions: there is no stretch of history, from a time to a time.

By contrast, time for Newton is very, very real—as it is for the modern world. Time is not for him or us a mere measure of what exists; it in some way preexists, and other things measure it, more or less accurately. We presuppose that things happen "in" it. This has consequences in every direction of our activities and our thoughts: theologically, for Newton, the world is stretched under high tension between the Creation and the Final Judgment; every event—not least, his own composition of the *Principia*—has its place on that divine time line, and its principal significance in relation to this ultimate historical framework. Human records are no longer a complication of tragic cycles but a tracing of developments through a succession of epochs in which the conditions of existence essentially change. The past is irreconcilable with the demands of the present, and a future which is likely to be very different is loaded with threat or promise.

Finally, or perhaps in the context of our present discussion, first of all, the art of mathematics is transformed by the idea of time. When Newton claims, in the Preface, that mechanics precedes geometry, he means that time is prior to Euclid's *Elements;* for the description of the line and the circle are processes in time, and mechanics, as the mathematics of motion, takes time as a primary mathematical object. Like the cosmos of Genesis, the mathematical objects which Euclid took as timeless are to be understood in their essence as products of a creation.

We shall see how Newton uses this new resource in shaping elementary propositions of his universal mechanics, when we turn in a moment to the mathematical lemmas with which he prefaces the *Principia*. For the moment, we may simply note the fact that the *Principia* will not only build its propositions on the presupposition of mathematical time as its material, but the work itself addresses and belongs to history. Newton is acutely aware of the place of his work in history—as we have seen, it is embedded in a sense of the Christian eschatology, marking as it does for Newton a stage of the revelation of truth to mankind which must be hard upon the threshold of the end of the world and the Second Coming. The fact that he can know what he has here come to know, and convey this to us in this work which tells us what it tells, is for Newton itself a mystery, an awe-inspiring development in our relation to God. It is an event: for Newton, God reveals himself historically in the course of these pages. And it will be equally in accord with this understanding to say that what the *Principia* teaches is finally a set of truths about history, and its method becomes a method for keeping time over the centuries and interpreting the evidences of the past, and thus an instrument for doing history. Thus, the work is wrapped in time: its propositions are made out of time, their unfolding is itself a crucial event in time, and their consequences teach us about time and history, on every scale. The *mathesis* of natural philosophy and, with it, of theology and history, is founded on the original *mathesis* of time.

What is this mathematical time which has thus taken central place in our minds? Newton is generous in stepping out of his text in reflective excursions he calls "scholia," to discuss with us the strange developments which he and we are witnessing together. He has these words of reflection on time:

Absolute, true, and mathematical time, of itself, and from its own nature, flows equably without relation to anything external, and by another name is called duration: relative, apparent, and common time, is some sensible and external (whether accurate or unequable) measure of duration by the means of motion, which is commonly used instead of true time; such as an hour, a day, a month, a year. (*Principia*, p. 6)

We note the formal reversal: Aristotle had said that time was the measure of motion; now Newton says, time exists in its own right, and motions become measures of it. Newton speaks here with the voice of authentic alchemy: it is exactly the alchemist's quest to find the true and philosophical existence beneath the common and deceiving appearances. The true mercury has something to do with appearances, but it does not normally manifest itself in the world of sensation and qualities—only by strategies of high art can it be induced to emerge, and when it does, we see evidences, but the unskilled eye does not see the underlying philosophical substance. True being is manifested in hieroglyphs, signatures which it is the work of philosophy to read. Newton is here doing with time exactly what as alchemist he was doing with mercury—and we see that in this way, he is laying the mathematical foundation for mechanics as rational alchemy.

Indeed, we see, as the discussion in this scholium unfolds, deep alchemy at work—and if we feel that the "time" Newton speaks of is the time of our modern laboratories, satellites, and stock markets, we may sense some working kinship with this Newtonian alchemy and, even in our jaded era, shudder in awe at the presence of mystery. Our clocks now derive their times from the motions of atoms, ranged from the immediacy of our laboratories to the edge of the cosmos. What do we imagine it is which coordinates the motions of myriad atoms throughout the stretch of nature? Our particle physicists speak earnestly

of events in the first seconds of the creation of the world of
the particles we know now: what coordinates those "seconds"
with today's "seconds"? Evidently, we, the true heirs of Newton,
believe in a time, flowing equably in itself, as an unbroken
foundation of being. Newtonian time is the first clue to our
modern alchemy!

Space

From time, Newton turns in this same scholium to space, about
which he finds a great deal more to say. His opening statement is, however, a parallel assertion of the existence of an
"absolute":

> Absolute space, in its own nature, without rela-
> tion to anything external, remains always similar and
> immovable. Relative space is some movable dimen-
> sion or measure of the absolute spaces; which our
> senses determine by its position to bodies; and which
> is commonly taken for immovable space.... (Ibid.)

In each of these statements, Newton's use of the term *nature*
is important, for it dispels any thought that these are mere
numbers, measures, or, in some nominalist sense, mere "concepts." No, it is clear, for Newton they are existent substances,
something very real at the root of the Creation. This is, again,
alchemy. We are not asked whether we wish (as the saying goes)
to "believe" in it, or whether we wish to "assume" or "suppose"
it. These dimensions are affirmed as beings, and the entire
work is built on the assumption of their reality. Nothing, large
or small, can move except in absolute space:

> All things are placed in time as to order of succession;
> and in space as to order of situation. It is from their
> essence or nature that they are places; and that the
> primary places of things should be movable, is absurd.
> (*Principia*, p. 8)

I take it that both time and space are understood by Newton to be true objects of intellectual intuition. That means, simply, that if we clear our minds of distractions, we can think about duration itself, and think about extension itself:

> …in philosophical disquisitions, we ought to abstract from our senses, and consider things themselves, distinct from what are only sensible measures of them. For it may be that there is no body really at rest, to which the places and motions of others may be referred. (Ibid.)

Is it not clear that in entering upon the *Principia* we are embarking on a work of *science* in the sense that we are occupying ourselves with matters which are knowable to mind and in no way contingent on empirical outcomes? Kant, it is true, will save the Newtonian enterprise in another sense, by claiming that these absolutes of time and space are no other than the modes of operation of our own minds, our ways of seeing the world (of *Anschauung*). Modern readers may be more comfortable with Kant's evasion than with Newton's clarity. But if we join Newton on his own terms, we see that time and space are now objects upon which certain knowledge can be built, as confidently as Euclid built on the intellectual intuition of a point, which could be alluded to but never constructed or presented to the senses in a drawing. The note of despair, suggested in the phrase "it may be that there is no body really at rest," reflects the heavy burden of the alchemist, who may never see emerge from the furnace secure evidence of the true philosophical mercury. But the faith of the alchemist appears in this scholium as well: "But we may distinguish," Newton immediately goes on to say, "rest and motion, absolute and relative, one from the other by their properties, causes, and effects." And with this begins a strategy for the extrication of truth from appearance which constitutes the balance of this long scholium and, in a larger sense, the work of the *Principia* as a whole. Just as he labored to refine the tangled mythology of traditional alchemy, Newton insists here on care in the use of terms, always with the aim of keeping reference honed to the underlying Being which is his unvarying object.

> … those violate the accuracy of language, which ought to be kept precise, who interpret these words for the measured quantities. Nor do those less defile the purity of mathematical and philosophical truths, who confound real quantities with their relations and sensible measures. (*Principia*, p. 11)

The fierce ethical tone of these lines speaks to the nobility of Newton's art and resonates with the romance of the alchemic quest: nothing less than truth and reality is to be its object. And we see alchemic practice at work in the method Newton adduces as the sure sign of absolute rotational motion—and hence, as well, of absolute rotational rest. He has found a procedure, a very alchemy, for the revelation of absolute rotational rest. In an ingenious experiment with an object which is no more than a rotating bucket of water, Newton finds in the curvature of the surface of the revolving water a signature of absolute rotational motion. What Newton is looking for is a sure sign in the realm of phenomena of our relation to something which is real, present, and knowable—though in the common course of the world, unseen and invisible. This surface curvature, like a telling color change in alchemy, becomes a revelation. As extended later in the scholium to the measure of tension in the "cord" connecting "globes," the same test bears on the absolute motion of planets, for the cord becomes the force of gravity; the globes, two heavenly bodies. By knowing the force between them, might we not, Newton is proposing, determine their absolute rotation and thereby detect the rotational posture of true and philosophical space? That would be the cosmic alchemy, to which Newton is devoting this work.

As we shall come to see, motions and forces are intimately related, true forces to true motions. Unless we can find our way to absolute space itself, the program of philosophy may be blocked. The purpose of Newton's philosophy, the Newtonian alchemy, is not simply to *know* what is true as theory but to *find out* true Being in the world. For Newton, theory is instrumental. The real and philosophical is to be sought in the world.

> And therefore as it is possible, that in the remote regions of the fixed stars, or perhaps far beyond them, there may be some body absolutely at rest; but impossible to know, from the position of bodies to one another in our regions, whether any of these do keep the same position to that remote body, it follows that absolute rest cannot be determined from the position of bodies in our regions. (*Principia*, pp. 8–9)

Yet the project of the *Principia* does not founder on this somber conclusion. For the real and true Newton will most want to discover is not time or space in themselves but an entity of far greater interest: that agency in the world he will introduce as "force." To reveal force, as we shall see, it is not necessary to find out absolute force directly.

The Lemmas

We have claimed that the style of the *Principia* is artfully matched to its purpose, and we have tried to give some anticipatory sense of what that purpose will be. Its aim, we have said in effect, is an intellectual intuition of true forms existent in the world. Nature is everywhere to accept and reward *mathesis*, while the *Principia* is to be shaped as the intellectual instrument which will make this new kind of seeing possible. That is the sense in which it is to provide the mathematical principles of a new philosophy. To get started on this sweeping project, Newton must first prepare certain appropriate mathematical methods. These are the new mathematics of motion, housed in the new absolute space and time. They have developed into the art we now call the "calculus," and which Newton, conceiving the project in terms of the mathematization of flowing or fluent quantities, called "fluxions."

Two Newtonian Problems

Problem 1: To find an instantaneous rate of change[11]

If we were to construct a Euclidean figure—let us say, a simple right triangle PDB (fig. 1a)—and then bring it into the Newtonian world, what world change might it undergo? From the purely relational figure of Euclid, existing timelessly and nowhere—for its own sake and according only to its own definition—it must now enter Newton's new conceptual time and space. Even as it sits quietly in the illustration, mathematical time is running, and the triangle is beginning, as Newton has said, to "endure." If we wish to, we can as well let it begin to alter in a regular way. We may for example let point B begin to move to the right at a regular rate. The triangle will begin to undergo what seems a Protean, fluent change, not only of place, but of shape. Newton's alchemy of motion demands that behind this confusion of appearances he must reveal clarity and constancy, ground on which to reason mathematically about such a blurred result.

---

[11] Brief summaries of Newton's mathematical work, his distinctive approach to the modern calculus, can be found in two useful collections: D. T. Whiteside, "Sources and Strengths of Newton's Early Mathematical Papers," in Robert Paleter, ed., *The Annus Mirabilis of Sir Isaac Newton 1666–1966*, pp. 69ff.; and Jon Pepper, "Newton's Mathematical Work," in John Fauvel et al., eds., *Let Newton Be!*, pp. 63ff.

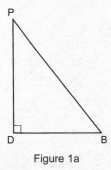
Figure 1a

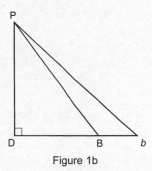
Figure 1b

In figure 1b, this motion has begun. Some interval of time has elapsed, and point B has moved to the right to a new position *b*. For that interesting moment at which the point was in its original position and the motion was just beginning, Newton coins a new and revealing term: he says the motion is "nascent"—in birth, just beginning. It is significant that Newton chooses the Latin verb *(nasci,* "to be born") which underlies the word *nature,* for this is the first step in the mathematization of what was not before mathematical, and it suggests the scope of the new mathematics. Nature is the domain of all things which undergo birth and death, beginnings and endings—to a first approximation, *nature* is simply Latin for Aristotle's term *physics;* the Greek *physis* is the realm of all that grows and fails, waxes and wanes. Both terms, Aristotelian physics and Newtonian nature, have in view the organic before the inorganic, and the motions of the psyche most of all. Such are the ambitions of this Newtonian triangle, the beginning as it is of the *mathesis* of all nature. Only in our own time, with advances such as those of psychoanalysis, neurophysiology, and information theory, are we beginning to glimpse the implications of this Newtonian mathematization of all nature.

As a consequence of the uniform growth of the base DB, the hypotenuse, of course, grows as well. Newton calls both DB and PB "fluent" quantities. The characteristic question for this

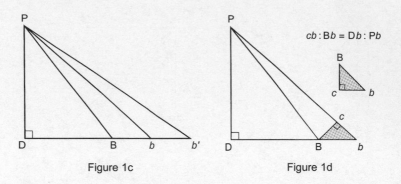

Figure 1c   Figure 1d

new geometry—Euclid's *Elements* set into motion—is, "What is the *rate of growth* of PB?" If we suppose that DB grows at a uniform rate, its length represents or measures time—it becomes a true and absolute clock in Newton's sense—and the rate of growth of PB can in turn be measured by comparing it with an increment B*b* of the uniform motion of the base. We have not at all solved our problem, however, for evidently this "rate" is as ever-changing, or fluent, as the two quantities of which it is composed. If at a later moment, depicted in figure 1c, we look at the triangle again, we see that the stretched-out hypotenuse P*b'* will be growing at a faster rate than it was earlier, as P*b*. How can Newton bring precision to discourse about a figure as intricately fluent as this?

The crucial problem before us, in specifying this rate, is to compare two changing quantities with one another. In this case, we want to compare the increments of the hypotenuse P*b* and the base D*b*. The little triangle B*cb* (fig. 1d) will serve to relate these two interesting increments. Here, *cb* is the increment of the hypotenuse, while B*b* is the increment of the base. To measure the fluxion, or rate of change, of the hypotenuse, we need the ratio of *cb* to B*b*. However, as that ratio is itself fluent, we have not by this device as yet escaped the morass. We need, it seems, to find its exact value at some precise moment—yet of course at any one moment, the little triangle, which is composed of *increments*, has no size, and we have nothing to look at or compare.

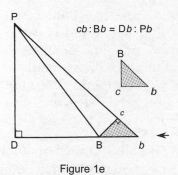
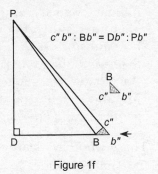

Figure 1e   Figure 1f

Newton solves the problem—and opens the door to a new world—in the following way. We may think of the motion as reversed, and let the increment B*b* shrink back to zero—this will be what Newton calls the *evanescent* motion, the image of the *nascent* in a mirror of time (figs. 1e–f). The required fluxion will be the ratio of the two quantities, *cb* and B*b,* just as they are vanishing—their ratio as they disappear. This would hardly seem to help, but Newton's method of following the behavior of these two disappearing magnitudes is characteristic of a method which will prove triumphant as the *Principia* unfolds. For he finds a way to keep these quantities visible to the eye, and to the eye of the mind, even as they vanish. He shows that the shaded triangle B*cb* is similar to the whole triangle PD*b*; this similarity is more evident if we redraw the little triangle right side up, as the small figure has done. Then as B*b* and the shaded triangle utterly vanish, the large triangle PD*b* will always make the shape of the vanishing triangle manifest to us. And finally, at the very point of vanishing, the shape of the disappearing triangle reveals itself as nothing but *the shape of the original triangle itself!* And the fluxion, namely, the ultimate ratio of *cb*:B*b* with which those quantities vanish, is the same as the ratio of the base to the hypotenuse of the original triangle. That is Newton's theorem, and the foundation stone of the new *mathesis* of all things that live and grow:

> There is a limit which the velocity at the end of the motion may attain, but not exceed. This is the

ultimate velocity. And there is the like limit in all quantities and proportions that begin and cease to be. And since such limits are certain and definite, to determine the same is a problem strictly geometrical. But whatever is geometrical we may use in determining and demonstrating any other thing that is also geometrical. (*Principia*, p. 39)

Those who have learned their calculus in modern courses, in the analytic mode by way of Descartes's algebraic symbols and the accompanying formal arguments, may appreciate the difference marked by Newton's intuitive diagrams and concrete representation of magnitudes as geometric figures. We should recognize, on the other hand, the universality of this method: the geometric lines may represent magnitudes from any domain of nature. In every case, the clue to Newton's method is insight: to substitute, for the mechanical operations of algebraic symbols, the invitation to intellectual vision embodied in a geometrical figure. And once again, Newton steps from behind the work to discuss with us a problem which has evidently occupied his own thoughts. We are speaking of the limiting ratio, as the quantities disappear. Should we admit that either the quantities have vanished, in which case there is nothing of which to take a ratio, or speak instead of very small infinitesimals and confess that we are not in fact taking the ratio as we had claimed, but taking a ratio of very small quantities just *before* they disappeared? To yield in either direction will mean abandonment of the enterprise. We must take the leap into complete continuity, the continuity of the line (and hence now, of motion on the line) assured by Euclid in Book X of the *Elements*.

It may also be objected, that if the ultimate ratios of evanescent quantities are given, their ultimate magnitudes will be also given: and so all quantities will consist of indivisibles, which is contrary to what Euclid has demonstrated concerning incommensurables, in the tenth book of his *Elements*. But this objection is founded on a false supposition. For those ultimate ratios with which quantities vanish are not truly the ratios of ultimate quantities, but limits towards which the ratios of quantities decreasing without limit do always converge; and to which they approach nearer than by any given difference, but never go beyond, nor in effect attain to, till the quantities are diminished *in infinitum*. (*Principia*, p. 39)

We see that the entrance into the new world of the *Principia* is by way of a wholly new insight concerning mathematics. We might say that at the root of this passage there lies a new view of the world itself, one in which time and space have become real, absolute, or mathematical in a wholly new way. As a consequence, motion has been transformed from what is inherently a blur, to be framed for thought only through the ingenuity of an Aristotelian philology, into a concept as precise and mathematical as Euclid's point, circle, or line.

Problem 2: Newton's Microscope[12]

If we are to implement a thoroughgoing *mathesis* of natural processes, we cannot be content to work with figures composed of straight lines. Everywhere we will be confronted with motions along arcs of curves, whether circular, elliptical, hyperbolic, or, within bounds, of arbitrary shape.

In figure 2a, we may think of point B moving along the arc B*c*A toward A. How can we prepare a *mathesis* adequate to its motion? To establish a ratio which will be of value in the study of motion along such curves, Newton draws the chord of that arc, AB, and then asks the ratio of the length of the

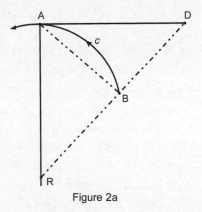

Figure 2a

---

[12] The technique of the "microscope" is introduced in Newton's Lemma 6.

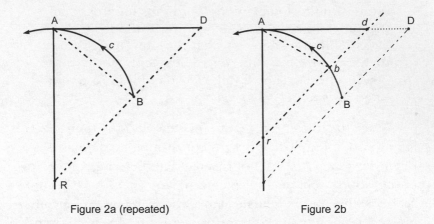

Figure 2a (repeated)   Figure 2b

arc A*c*B to that of its chord AB. To get a further ratio, a second measure of the same motion, Newton next draws the
line RD in any arbitrary direction, so that it cuts off a definite
length along the tangent drawn to the arc at A. We are invited
to watch the evolution of these ratios as B moves toward A.

As B moves to *b,* and the arc thus shortens, the line *rd* is
always to be drawn parallel to its original position, so that a
definite rule is established for cutting off the length along the
tangent (fig. 2b). We are now tracking the evanescent motion
to get *two* fluxions: the ratio of the arc to its chord. and at
the same time the ratio of the arc to its tangent. The fluxions
in question will be the ultimate ratios as *b'* finally reaches A,
and once again all the quantities we are interested in frustrat-
ingly vanish. Again, Newton will rescue the situation by using a
brilliant rhetorical device to make the vanishing ratios remain
visible to the eyes of our minds.

Returning to the first position of arc, chord, and tangent,
Newton now sets up what is known to tradition as his "microscope"—an enlarged figure, with arc A*mk,* which is similar to the
one in which we are interested (fig. 2c). The arcs A*cb* and A*mk*
have exactly the same shape, and differ only in size—exactly as
if A*cb* had been magnified in a photographic enlarger.

In figure 2d the arc, which we may now designate A*b',* has
become very small, and in figure 2e, as A*b'',* one may think of

it as evanescent, on the verge
of vanishing. Yet Newton has
contrived to keep it fully visible: the shapes and relative
sizes of arc, chord, and tangent are precisely exhibited to
the eye and the mind in their
enlarged counterparts, the
power of the rhetorical microscope having increased in just
such proportion as always to
keep the size of the enlarged
figure constant, though at the
same time it takes the changing shape of the vanishing arc
segment. Since there is presumably no limit to the power
of a conceptual microscope,
we see clearly the theorem
Newton is placing before us.
In figure 2e, as the arc, its
chord, and its tangent all vanish like a trio of Cheshire cats,
their magnified images will be
preserved and will come to
coincidence in the limit. The
large arc A$m''k''$, its chord Ak'',
and the tangent AD *coalesce* as
the arc Ab'' vanishes. In figure
2e we glimpse them as they
disappear over the conceptual
horizon, at once vanishing
from view and becoming fully
manifest in their true relationship as nascent quantities.

Once again, the question
of fluxions has been answered:

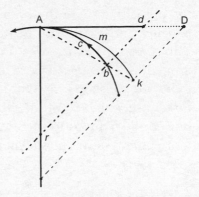

Figure 2c

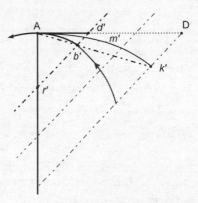

Figure 2d

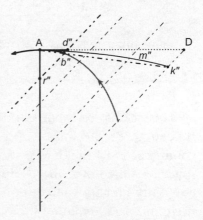

Figure 2e

the ultimate ratio of the evanescent arc to its chord is that of equality, and so likewise is the ratio of the arc to its tangent. This amazing fact licenses much of the work which must be done in the *Principia,* where arcs of trajectories and orbits are to be our constant fare. Within the limits of reasonable curvatures with which Newton will be working, curvilinear is thus universally reduced to rectilinear. No transmutation could be more potent in this new alchemy of all nature.

Definitions: Matter and Force

We have seen instruments prepared by which the world of phenomena is to be revealed in a new order of clarity as a mathematical system. Newton has a distinct vision of the structure of this system, the groundwork for which is firmly laid in a set of Definitions and Laws of Motion at the outset of the work.

Matter

Underlying the entire natural world is something Newton calls "matter." He does not in fact define it, but rather, in accord with the concept of the *Principia* as the mathematical principles of the world, defines the *quantity* of matter. But the form of this definition reveals what he has in mind:

> The quantity [mass] of matter is the measure of the same, arising from its density and bulk conjointly. (*Principia*, p. 1)

This seems innocent enough, but in fact it bears careful reflection. By the "mass" of matter he means the amount or volume of it. By the density, he means in effect the fraction of space occupied by matter. Thus the product "density and bulk conjointly" means "percent of space occupied by matter times the total space = volume [amount, or *mass*] of matter." What is presupposed is that matter is utterly simple, compact, and one. There are no kinds of matter: it is not that copper is one kind of matter (with one density) and lead another (with a greater density). This is the ultimate matter of the alchemists, which might have been fetched out, had Newton felt a need to discuss the question, as the *true and philosophical* matter, the analogue of the true and philosophical time and space. This is the alchemist's vision he is invoking, seeing through to that ultimate and invisible matter which finally underlies all commonplace distinctions. This ultimate matter has no properties—no color, no flavor; it is not magnetic or gravitational. Its character is that it is absolutely passive. It does not belong to the idea of matter to be hard; Newton has in fact come to believe that the matter of the world is organized in totally hard, permanent particles—atoms—but he equally admits as an idea conformable to these first principles a matter which, though otherwise identical to atomic matter and equally dense, is perfectly fluid. A great deal of the *Principia,* its second book, is devoted to serious discussion of that possible fluid matter.

This Newtonian matter has one, single principle of motion and rest. That principle, called *inertia* (literally, lacking any art or skill), expresses precisely the passivity of matter and gives rise in interactions with other bodies to something called the *vis inertiae,* the power or force of inertia. Newton calls this a *passive force.* We cannot look to Newtonian matter to initiate any motion in the cosmos: the agent in any action must always be sought elsewhere.

This is the whole story of Newtonian matter. If there were no other principle in the natural world than matter, we would have to derive the reconstituted mathematical cosmos from this concept alone: matter moving in space and time, reacting to chance encounters with other matter, but with no agent principle in the world at all. It would not be a cosmos: it would be a dark and dead world, perhaps that of Lucretius, if the matter were fixed in atoms, or that of Descartes, if the matter were continuous and fluid. The proposition that the world is of this sort is what Newton calls materialism, and the *Principia* can best be understood as a dialectical work designed to refute the idea that the world is of this sort, and to reveal and demonstrate the richer alternative. Again, we see the power of the Newtonian alchemy: by insisting on the purity of the idea of matter, as absolutely inert and without life, Newton sharpens the idea of its opposite, which the alchemists call *spirit*. That is, his vision of God as totally governing and the cause of all act demands as counterpart the idea of matter which contributes nothing. In that sense, the concept of matter as undifferentiated and inert is the other face of the idea of God.

Who is the antagonist in Newton's dialectical enterprise? Not Lucretius, for Lucretius has not brought matter into the domain of mathematics, and materialism becomes threatening only when it becomes mathematical and invites the mind. The mathematical Lucretius, and hence the serious adversary, is Descartes, who fills the natural world with a continuous and totally intelligible fluid matter. As we have seen, Newton has learned the very conception of natural philosophy as mathematical analysis from Descartes. A wave of enthusiasm among thinking persons has greeted Descartes's *mechanical philosophy*, for it admits the possibility of explanation—thoroughgoing mechanical explanation, in mathematically consistent terms. Perhaps the universe is nothing more than a mathematical clockwork, which could be understood in mechanical terms throughout: Newton knows well, at firsthand, the lure of that possibility. All the more urgent, then, that the distinction now be drawn sharply: Newton directs the *Principia* to the purgation of the fundamental error in Descartes's account of the physical world. The second book will take Descartes's principles, calculate the consequences, and undertake to show that the universe which would result is far from the one in which we find ourselves. Kepler's laws of planetary motion would not hold, and the heavenly bodies would dissipate their motions and halt in their courses.

Active Force

The new element that Newton introduces into the world, in every way the opposite of matter, is an active principle which he calls the *vis motrix*—the "moving power," usually translated "motive force." This is Newton's counterpart in the *Principia* to the *spirit* of the alchemists. It is invisible, is intangible, freely penetrates matter, and everywhere works with vitality and sure signs of intelligence to accomplish the design of the cosmos. It does this by moving always according to perfect mathematical law—"law" which Newton understands first of all in the political sense, as God's ordinance, the established vehicle of God's will in the governance of the world. The *Principia* in the formal development of its mathematical principles does not address such philosophical issues, but it makes way for living force or agency by way of the *vis motrix*. As it turns out in the philosophy which Newton builds upon the *Principia,* there are several such active principles in the world: Newton speaks of the cause of gravity, of the force of cohesion, which causes bodies to form up and remain intact, and the inner principle of chemical, vital activity, which alchemy calls fermentation. These will be the true causes of the motions of the world we know, animate as well as inorganic. They will be the primary objects of philosophy, and the counterparts to Aristotle's *physis.* It is the goal of philosophy to find these first principles of all life and motion in their universality, and to trace the ways in which they cooperate to construct the living cosmos. They appear in the *Principia* as distinct measures of the *vis motrix.*

The Laws of Motion

Newton formulates three postulates which govern all motion in the world. He might have called them "principles" of motion, but he calls them "laws." It is completely consistent with his view of the world that he do so, for as we have seen, matter,

as Newton has distilled the concept, has no propensity to new
motion and must be commanded if it is to be moved. Action
is utterly external to matter, and thus the initiation of motion
comes through the externality of edict or law, which has no
coloration of persuasion. This stark character of the laws of
motion rightly suggests the spirit of the Old Testament: God
commands and obedience is utter. Between God who is all act
and matter which is inert, the mode of communication is law,
and the resulting order is one of dominion.

It is possible to look at the same world in more than one
way. Leibniz, operating from a different sense of the nature
of things, is able to use the Aristotelian paradigm of potentiality and actuality, and to see in a material system at rest the
potential for motion—potential energy which will be actual-
ized by its own nature as living force. The difference shows
itself in many ways, leaving its trace in virtually every way we
speak of the phenomenal world. Mind approaches the world
with mythic predispositions and finds ways to tell a story which
reads in ways it needs to hear. What measures do we take of
motion and force? Newton tells us that the quantity of motion
is the product of the matter and motion conjointly—mass and
velocity, yielding, we say, momentum. Leibniz, thinking otherwise and perhaps listening to the voice of a different god, takes
as measure of motion the kinetic energy, an approach which,
followed consistently, traces out an utterly different physics. All
Newton's reckoning of his laws of force will be in terms of this
quantity on which he has fixed his attention. The first law preserves that quantity:

> Every body continues in its state of rest, or of uniform
> motion in a right line, unless it is compelled to change
> that state by forces impressed upon it. (*Principia*,
> p. 13)

The second demands that the force be proportional to the
change of that chosen quantity:

The change of motion is proportional to the motive force impressed; and is made in the direction of the right line in which that force is impressed. (Ibid.)

The third asserts that in interactions the exchange of quantities of motion will balance, with the result that in an interaction no change at all of the quantity of motion occurs:

To every action there is always opposed an equal reaction: or, the mutual actions of two bodies upon each other are always equal, and directed to contrary parts. (Ibid.)

What is the significance of this quantity which has become so fundamental? Consider a chamber full of particles moving at random. For Newton, their total quantity of motion is nothing, for the vectors sum to zero. For Leibniz, the chamber is full of living force, and that becomes our understanding of the temperature of the system: thermodynamics and many of the finest insights of modern science flow from Leibniz, not from Newton. Although it is possible to find hints of kinetic and potential energy measures in the *Principia,* on the whole Newton, thinking of force in terms of the directed blow, is limited to a vector physics. The alternative insights of the scalar physics of Leibniz, in which processes of a system unfold spontaneously, from energy stores within, are closed to him. Each physics paints its own, very distinct myth of the world. Newton's laws are blinders as much as they are lights—to tell one story whole, we may have to be blind to others. Today we imagine we read and reconcile both stories, but we no doubt tell stories sealed with blinders of our own.

I have suggested that Newton's laws of motion breathe the air of the Old Testament, and I suspect that all we have just said tends to confirm that view. But this is only the beginning of Newton's tale, as the Old Testament is the beginning of Scripture and speaks of the beginning of time. Like the

Scriptures themselves, the *Principia* moves by prophetic path-
ways through darkness toward new light, and we might sense
that Newton's vivid sense of a new relation to God is at stake as
that process unfolds.

The Hieroglyph for Force[13]

The laws of motion have taught us that if we are to read the
book of nature, we must seek out the forces which are operating
as the causes of all life and motion. How from the phenomena
are we to find those forces—how can we penetrate the appearances, to find the true principles which lie beneath? This is the
great alchemy, whether of macrocosm or microcosm, heaven
or earth: to seek the true and philosophical principles which
underlie the distracting and confusing surface of events. If, as
it seems, it is good counsel to regard nature as a text, our task
is indeed one of interpretation: we must learn to read a book
written in the patterns of nature. But are these phenomena
legible characters? Newton now pursues the proposal that they
are, and that patterns can be found in them which are intelligible. These can be taken as signs or indices to causes, active
forces, which lie hidden. He shapes a fundamental geometric
instrument of a special sort, at once image of a structure in
nature and index to underlying truths. Newton worked with
powerful signs and symbols in his studies of alchemy, among
them those in the form of hieroglyphs. I suggest that we may
think of the geometric constructions of the *Principia*, combining representation with coded significance, as "hieroglyphs"
in this sense. Let us see how Newton puts his new geometric
method to use in this manner, to shape what we may think of
as a hieroglyph for force.

---

13 According to Newton, "the language of the Prophets, being
Hieroglyphical, had affinity with that of the Egyptian priests and
Eastern wise men, and therefore was anciently much better under-
stood in the East than it is now in the West" (McLachlan, *Theological
Manuscripts,* p. 120).

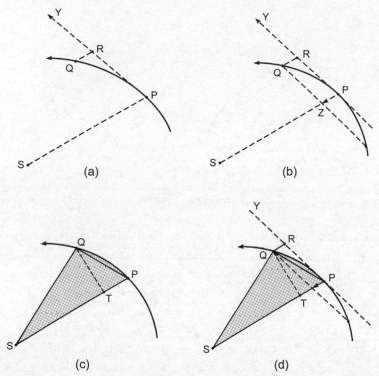

Figure 3. This figure, which we are calling Newton's "hieroglyph for force," is in effect a movable geometrical instrument, which can be used wherever needed to take the measure of a force deflecting the trajectory of a body toward a fixed center. Here S is that center of force, while P is the point whose trajectory is undergoing deflection. We may think of S and P as "sun" and "planet." In the absence of the central force, the point would move in a given time interval in a straight line toward R. Instead, however, the force toward S gives rise to the deflection QR, which thus becomes our initial measure of the force. Since the same force will give rise to greater deflections if it acts for longer times, we must contrive to take the time of action into account. As figures 4 and 5 will explain, the triangle PSQ becomes that measure.

According to Newton's First Law of Motion, or indeed to his very idea of matter, a body moving at point P in figure 3a with no force acting on it will follow a straight line at uniform velocity—that is, the body at P will follow the straight line PY. If a force acts upon it in the direction S, however, the body will be deflected according to the Second Law from the straight line and will as a result trace some arc such as PQ. Newton now constructs a strategic geometric instrument by which this force can be found out and measured. If the time

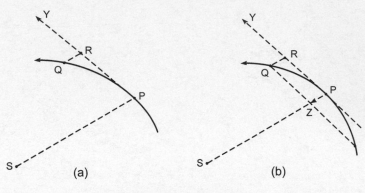

From Figure 3 (repeated)

did not have to be allowed for, the deflection itself, measured by the line RQ, would suffice as a measure of the force. In figure 3b, this deflection is mirrored in the line PZ, for in the parallelogram QRPZ, the opposite sides RQ and PZ are equal. PZ is here drawn as an arrow pointing in the direction of the center of force. This arrow drawn between an arc and its chord in the direction of a center of force and equal to the deflection caused by the force, Newton calls the *sagitta* of the arc. We recognize it as a vector, with direction and magnitude.

We face an important complication, however, for the same force acting for a longer time will produce a greater deflection. Evidently, our sagitta measuring the force must in some way be discounted according to the length of time during which it has been produced. An initial step, then, must be to get some measure, within the diagram itself, of the time which has elapsed as the arc PQ was swept out.

The solution to this problem is suggested by Kepler's Law of Areas: if a moving radius is drawn from a planet to the sun, that radius will in a given time sweep out equal areas in equal times (fig. 4). Cumulatively, this implies that if the times are unequal, this body will sweep out areas proportional to the times (fig. 5). If we generalize the rule from the planet and sun to any body moving under a central force, we may thus take the area swept out in the figure as geometric measure of the time.

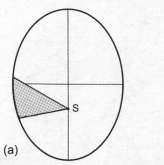
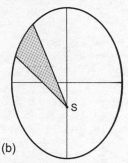

(a)   (b)

Figure 4. Kepler's law of areas asserts that, in the motion of a planet about the sun, equal areas are swept out in equal times. Newton's Proposition 1, Book I, has extended this to apply to any body moving under the action of a central force alone. Here the two shaded sectors are swept out by such a body in equal times, the motion speeding up at (a) to compensate for the shorter radius.

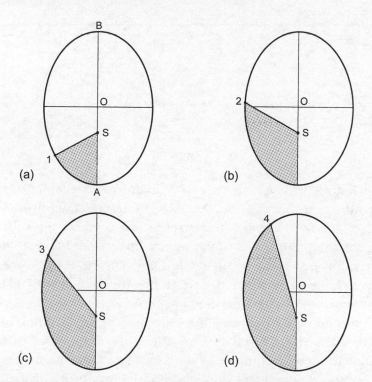

Figure 5. Since equal areas are swept out in equal times, as a body moves under a central force the sector it defines will increase in cumulative area in direct proportion to the time elapsed. Here, as the body moves in equal time intervals through positions 1, 2, 3, and 4, the area of the sector similarly increases by equal amounts. The problem of constructing an almanac of the body's position in the heavens, becomes the geometrical problem of finding the areas of these sectors.

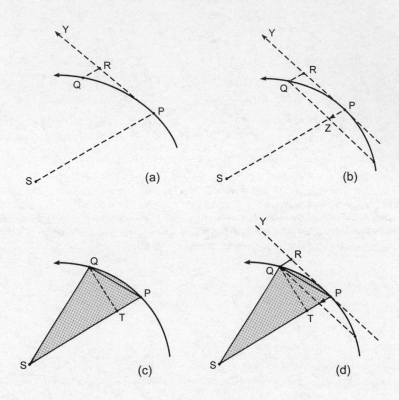

Figure 3 (repeated)

In figure 3c, then, the area PSQ represents the time in which the body has traversed the arc PQ. In effect, we are setting up geometric figures to represent, or serve as signs for, the physical quantities in which we are interested: here, while we have proposed the sagitta as sign or measure of the force, the area PSQ becomes a figurative clock. And we may now take advantage of the license granted by our earlier investigation of fluxions and equate the arc PQ to the straight line which is its chord. We have only to focus our attention on the arc PQ in its nascent stage, zeroing in by the same token on the precise force at the very moment when the body is at point P.

Combining the sagitta measuring the deflection with the area measuring the time in one complex image (fig. 3d) will give us the ingredients of the true measure of the force. To effect this combination, Newton draws upon one further

insight: with respect to its radial motion, body P is in effect
falling freely toward its center S. But with that recognition, we
see that we can use a theorem of Galileo's to the effect that a
falling body traverses distances proportional to the squares of
the elapsed times. The sagitta will in general be of a length
proportional to the square of the time in which P has traced
out arc PQ.

From these ingredients arises the formula which unlocks the
meaning of Newton's complex sign. The sagitta PZ equal to QR
increases in proportion to the square of the time, and hence
for a given force its length must be discounted by the amount
of the time: we must divide the quantity QR by the square of
our measure of the time. But the area PSQ measures the time,
and in turn, in the limit, taking the chord PQ for the arc, this
area measuring the time will be that of the triangle PSQ. If we
drop the perpendicular QT, the area of that triangle will be
one-half its base times its height, or proportional to (QT·SP).
Finally, taking QR as measuring the deflection but discount-
ing it in proportion to the square of the area (QT·SP), we get
Newton's formula for reading his sign:

$$f \propto \frac{QR}{(QT \cdot SP)^2}$$

Figure 3d appears in Newton's text in the form of figure 6,
below. It becomes a powerful sign for the interpretation of

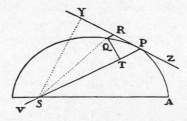

Figure 6. This figure illustrates Newton's Proposition 6, Book I, measuring
the centripetal force in the case of a body which "revolves in any orbit about
an immovable center, and in the least time describes any arc just then
nascent." His words remind us that these relations hold only for "nascent"
quantities, i.e., in the limit, at an instant.

phenomena on any scale, anywhere in the universe, and for finding out forces of any kind. It is thus a universal analytic instrument for finding out causes, not merely qualitatively, but with quantitative precision. We may think of it as belonging to the rhetoric of interpretation, the alchemist's dream, converting the phenomena of nature into a readable text, and making it possible to penetrate the surface of appearances in order to find out the true and philosophical forces at work beneath them. We are reminded that rhetoric is in one sense the art by which we grasp one language in terms of another; it is thus a work of translation, here connecting appearances with their underlying and governing forces, which we can only see in this way—nature is unintelligible without interpretation.

Laws of Force

We have already glimpsed the structure of the cosmos as Newton will depict it. Thus far, however, we have looked only at the definitions, lemmas, laws of motion, and the first of the propositions of the *Principia*. We have developed powerful instruments for the geometric analysis of the world but have not yet put those instruments to actual use. We have the tools of a new literacy, a grammar of nature, but have not yet approached the text.

To follow Newton now, we need to have a better sense of the organization of the *Principia*. As we have seen, the stance of the work is entirely instrumental: even as we now put these new mathematical instruments to work in actual analysis, the results remain hypothetical. The entire first and second books of the *Principia* are preliminary, wrapped in the syntax of conditionality: "If the phenomena are of a certain sort, then the consequence will be …." It is only in the third book, "The System of the World," that Newton formally turns from what we might think of as the book of all possible worlds to the actual phenomena of the world we live in. At that

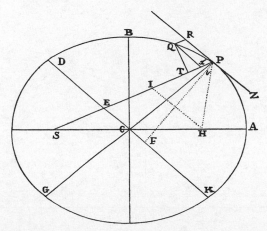

Figure 7. Newton's figure for Proposition 11, Book I, in which a body moves under the action of a force varying inversely with the square of its distance from a center. This proves to be the case when the body P moves in an ellipse about a focus S. The force is least at aphelion A and greatest at perihelion, at the opposite end of the major axis.

point hypothesis converts to demonstration, mathematics to philosophy.

We turn to Newton's fundamental Proposition 11, which, though hypothetical, is keyed to the phenomena which will later emerge as our own:

> If a body revolves in an ellipse; it is required to find the law of the centripetal force tending to the focus of the ellipse. (*Principia*, p. 56)

His drawing is reproduced as our figure 7. We see the ellipse, with one focus at S (the letter suggesting what may emerge as the sun) and a body at P (to emerge in Book III as the planet), and we recognize as well the hieroglyph, our instrument of analysis, superimposed on the figure. The force is measured by the formula we have prepared, its elements being QR, QT, and PS. To solve this problem, Newton has only to weave these elements into the geometry of the ellipse as taught by Apollonius. We will not track those steps, but the outcome is direct: "the centripetal force is inversely as … SP^2, that is, *inversely as the*

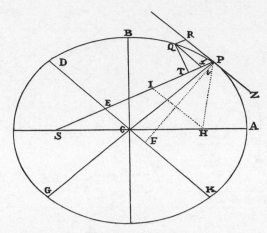

Figure 7 (repeated)

square of the distance SP." As the body traces the circumference of the ellipse, its distance from the center of force at the focus will vary, being greatest when the body is at aphelion A and least when the body arrives at perihelion, at the opposite end of the axis AS. The force being greatest as the body approaches the center and least as it removes, it will be most attracted at perihelion and least at aphelion, in the proportion of the *squares* of the corresponding distances—i.e., very much more attracted at perihelion, and very much less at aphelion.

We may allow ourselves to anticipate the turn of the text in Book III and thus to recognize that here we have the phenomena of our own world. In application to the phenomena of our heavens, this analysis will reveal that the planets are bound to the sun, our moon to earth, and Jupiter's moons to Jupiter, by a force which obeys this inverse square law.

Again, Newton's account is in terms of "law," but now that concept is being applied in a somewhat different sense. The former laws, "laws of motion," applied to all matter and any motive force, by virtue we might say of the very notions Newton has arrived at of matter and force *per se.* But this law will apply to only certain of the forces Newton is searching out in nature, such as the one he comes to call "gravity," and it speaks of the distribution of this active principle over space. Other forces, the life-force of "fermentation" for example, or that of cohesion, or of magnetism, may be distributed according to other laws—the elastic force, for example, acting to repel rather than to attract. These are, then, laws of force in a different sense, forms of the operations of the various modes of agency. In fact, Newton goes on to generalize his analysis in this section to consider a continuous range of all possible force laws, attractive and repelling, bonding more tightly or more loosely, which might go into the making of possible universes of very different sorts.

To put this concept in terms of the theology which for Newton is its ultimate context, these forces are the modes of God's very governance of our world. What is emerging is the recognition that *mathesis* is God's chosen mode of action, from which it follows that God's activity is in some way intelligible to our minds. Gravity as an inverse square law holds each planet in a balanced orbit: as Newton's propositions in this book show, a tighter force law (varying inversely by a power greater than the square) would have pulled it to the center, a looser force law would have lost it to infinity. I suspect that the tone of Newton's contemplation of this intellectual order is not one of intellectual arrogance—to reduce the divine plan to the level of human strategy—but rather, immense awe at the immanent presence of a mystery. The world is revealed as a mathematical construction perfect and harmonious as the regular solids with which Euclid brings the *Elements* to conclusion. There is no *mechanism* for these divine actions: they are pure acts. Not only are they constantly and perfectly proportioned, but we can see by ranging over the options, as Newton's broad analysis permits us to do, that this proportion has been so chosen that the whole can form a lasting harmony. To watch the planets move under the guidance of such actions must have meant for Newton presence at an immanent, divine mystery. It might be so for any of us, if long habit had not made these matters seem banal, topics for textbooks.

Newton regularly exhibits a breathtaking ability to achieve
an overview of a subject which it would not occur to lesser
minds even to propose. That is, the Newtonian projects often
exceed in their scope any reasonable level of anticipation. Such
is the case with the vision he gives us in waves of propositions
which follow upon the first demonstrations of motion under
the action of a central force. We cannot do more, here, than
note the great benchmarks of this enterprise—and then later,
in due course, observe how these elements fit into Newton's
still grander plan, as these still only mathematical instruments
play their roles in the construction of the System of the World,
in Book III. Once again, although it is crucial to Newton's
design that these propositions preserve their integrity as strictly
mathematical demonstrations, it is clear that they are everywhere shaped to the work to which they will be put when the
time arrives, in Book III, for the turn to the phenomena and
the interpretation of the world as it indeed is.

We have arrived, in Proposition 11, at that point at which
Newton has analyzed the case of motion in an elliptical
orbit under the action of a central force directed to one of
the foci and shown that in this case the law of force will be
that of the inverse square. In application to the motion of a
planet, this yields a dynamic account of the first of Kepler's
laws, that the geometric figure of the orbit of a planet is the
ellipse. Newton had already shown in his first proposition
that a planet's adherence to Kepler's Second Law, the law of
areas, interprets dynamically as a sure sign that the body is
moving under the influence of some central force, located at
the point from which the areas are measured. What, though,
of Kepler's Third Law, that the periods of the planets are
to one another as the $3/2$ power of the ratio of their radii?
Reasoning by way of the known relations of the areas of
ellipses, together with the known relation of area to time,
Newton makes quick work, in Proposition 14, of this final law
of Kepler's, as well. Kepler's laws, as laws of phenomena, have
thus been penetrated to yield their underlying secret—that

they are summarizing the phenomena under the action of an agency, centered at the focus of the ellipse, and obeying the law of the inverse square.

Newton now poses what becomes, in the later discussions, God's problem: how to launch a body in relation to a given center of force with just such a combination of position, initial direction, and speed as to insure an orbit of precisely a required size and shape? Using the combined principles of geometry and dynamics already at hand, Newton is able to solve this well in Proposition 17. In a celebrated later correspondence on the theological significance of the *Principia*, as he weighs the consequences of all the options which might be considered as possible beginnings for our solar system, Newton concludes, with an eye to this proposition especially, that the harmony of its design

> argues that Cause to be not blind and fortuitious, but very well skilled in Mechanicks and Geometry.[14]

At this point, Newton might well have rested, but far from it, he embarks now on one of his most impressive adventures of mind. He invents a sensitive hieroglyph for detecting any slight departures from the inverse square force law which he has just finished establishing; he shows that to the extent the law of force is disturbed, the axis of the ellipse will wander. Further, reversing this principle, he utilizes the moving ellipse constructively to generate his exploration of the spectrum of orbits which arises under the full range of possible algebraic force laws. This device of the moving ellipse reveals its importance, since Newton goes on to show that nowhere in the cosmos will force laws operate in isolation: the very coherence of the universe requires that all its forces be disturbed, that no ellipse rests.

---

[14] The correspondence was with the theologian Richard Bentley. See I. B. Cohen, ed., *Isaac Newton's Papers & Letters on Natural Philosophy* (Cambridge, Mass.: Harvard University Press, 1978), 2nd ed., p. 287.

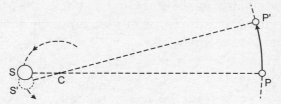

Figure 8. In earlier propositions, Newton has assumed a central body about which another revolves. Thus in Proposition 58, Book I, we find that if a body S constrains P to an orbit, S must orbit as well in reaction. All that remains fixed in space is the empty point C, the center of mass of the pair.

The Three-Body problem

Newton's next major step is momentous. He begins a motion from a fictitious world of mathematical orbits computed about fixed centers to one of a very different sort, and far more realistic. What makes this inadequate is that nowhere in the cosmos is a center fixed: nowhere is there terra firma in which to anchor a central body. Newton's Third Law of Motion assures that forces are interactive, and with that recognition comes a wave of fundamental consequences: a central force is not a one-way street but in its very conception involves an interchange and a dynamic balance. Every body is both mover and moved; motions must mirror one another. If the first of the great astronomical revolutions required that the earth must move, the second, entailed as soon as the Third Law is uttered, is that the sun must move as well (fig. 8). The sun will not be the center of the solar system: it too must orbit. At the erstwhile center there will now be nothing: only the mathematical balancing point, the center of mass, about which all bodies of the connected system make their one, configured dance. It is this recognition, fraught as we shall see with consequence, that Newton has in view when he speaks of "system"—and names his third book "The System of the World."

The first step into this new world is very easy; Proposition 57 (of Book I) shows us that orbits will be geometric mirrors of one another. To the earth's large ellipse about the center of force will answer another, the sun's own orbit, small in

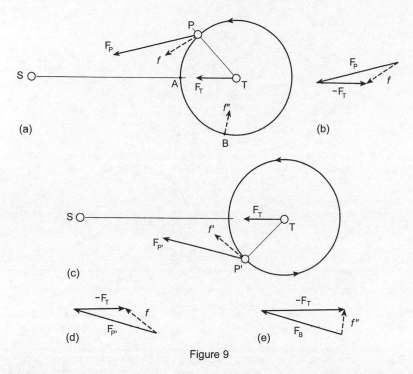

Figure 9

proportion as the sun's mass is great. Proposition 58 continues the inquiry, to address the question of time: the earth's year, it develops, will be lengthened by this consideration; there is a relaxation of the central force as the center now yields. In the same way, the earth yields to the moon, and the month is the longer, as the earth's hold slackens.

But the logic of this line of thought carries us beyond the interaction of two bodies to the consequence that there will always be third bodies troubling the motion of any two. The moon's motion about the earth will be affected as well by the sun's force. If the sun attracted both earth and moon alike, the two would together fall toward the sun, and their mutual motion, the monthly course, would be unaffected. It is the *difference* in the attractions, as the moon's distance from the sun becomes alternately more and less than the earth's, which affects the relation of moon to earth.

Newton addresses this ominous problem in Proposition 66, the diagram for which is adapted as our figure 9. Thinking in

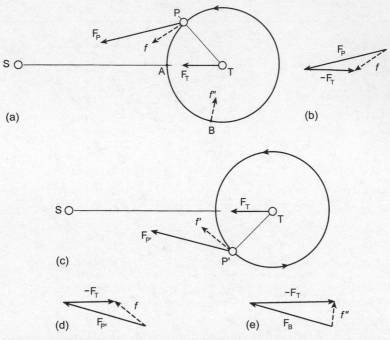

Figure 9 (repeated). When a third body is introduced into the situation of figure 8, the orbits become distorted. If P is the moon and T the earth, the sun S will have a disrupting effect on the moon's orbit. The moon at P is closer than the earth to the sun, so vector F_P measuring the sun's force on it is greater than F_T measuring that of the sun on the earth. The vector triangle at (b) determines this difference as f, which when drawn at (a) is revealed as accelerating the moon and stretching the radius. In (c), the moon at P' is retarded, and at B, as the vector diagram (e) reveals, the radius is shortened.

this way of the sun's interference with our moon's motions, Newton labels the figure S (the sun) and T (the earth), while P is the point at which the moon is located at any moment in its course. What must be reckoned is the effect of the differential force on the moon, indicated in our figure by the vector f, and at this point even Newton has met his match: our diagram depicts the infamous "three-body" problem of mechanics, which is not solvable in closed form. That is, though approximations can be given as to the effects of the third body—of the sun on the moon's motion about the earth—no single formula exists which asserts the result in a single mathematical statement. We can readily see that when the moon is at position P in its passage toward A, it will be accelerated in its course by the disturbing force *f,* while at P′ it will be retarded. But the "orbit" is not a fixed track. It too will undergo distortion, and we see that it is not strictly true, as Kepler asserted and Newton seemed to have demonstrated, that the paths of the planets will be ellipses. For here the ellipse will be bent out of shape, bent toward the sun and away from the earth when the moon is at A, and toward the earth when the moon is at B. Everything in Newton's cosmos is bent out of shape.

We are speaking of the case of the moon's motion about the earth, and indeed that is one of the first concerns, as Newton's labeling of the figure for the proposition suggests. But it is only the beginning of the unfolding consequences. All of the orbits and times are similarly affected—every motion will affect and disturb every other. Earth in its motion about the sun will thus reflect the positions, and motions, of all the other planets, as well as the moon's motion about the earth— so that distortions and corrections must be made to the entire system. They are all in the same way, incalculable—capable, in principle, of being approximated insofar as one can find strategies for carrying out the series of arithmetic corrections necessary, but not capable of solution in good, closed form. The order and beauty of the mathematical figures and the regular times is forever destroyed. Newton spreads before us this dark revelation, in the proposition itself, whose terms become obscure: indeterminate weighings of more and less replace the precision of strict magnitude and ratio, the language has become tangled and intricate:

> If three bodies, whose forces decrease as the square of the distances, attract each other; and the accelerative attractions of any two towards the third be between themselves inversely as the squares of the distances; and the two least [P, S] revolve about the greatest [T]: I say, that the interior of the two revolving bodies [P] will, by radii drawn to the innermost and greatest [T], describe round that body areas more proportional to the times, and a figure more approaching to that of

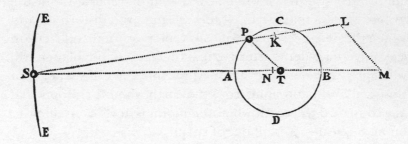

Newton's diagram for *Principia*, Book I, Prop. 66

an ellipse having its focus in the point of intersection of the radii, if that great body be agitated by those attractions, than it would do if that great body were not attracted at all by the lesser, but remained at rest; or than it would do if that great body were very much more or very much less attracted, or very much more or very much less agitated, by the attractions. (*Principia*, p. 173)

(Here P and S may be treated as "lesser" bodies, in the sense that, being remote, the influence of the sun, though important, is small with respect to the moon's motion about the earth.) The corollaries, fraught with consequence, continue in the same mode (see Newton's figure at top of page):

COR. II. In a system of three bodies, T, P, S, if the accelerative attractions of any two of them towards a third be to each other inversely as the squares of the distances, the body P, by the radius PT, will describe its area about the body T swifter near the conjunction A and the opposition B than it will near the quadratures C and D....

COR. IV. The orbit of the body P, other things remaining the same, is more curved at the quadratures than at the conjunction and opposition.... (*Principia*, pp. 176–177)

We are mired in unpleasant discourse: here is a language of confusion, not of clarity, or of the science of which Newton has proved himself master. Yet it is not some special case, a difficulty which presents itself in certain unfortunate circumstances. It is the universal case: never does a second body orbit about a first without the interfering influence of a third body—and indeed, this sequence of complications never stops with a third, but we are confronted now with the interfering effects of every body on all bodies. The problem of perturbations is universal, endless in its implications and devastating to the project of strict scientific computation. The system of the world is not, finally, knowable: it escapes true knowing, or at least, human knowing. Newton, as we have suggested, confronts this problem bravely, but he has no recourse. There is no solution. There still, in modern science, is no solution. Computers make their ways effectively enough through the morass of complications, but they yield outcomes, not knowledge.

In classic terms, the outcome must be seen as tragic. If Newton's goal were to know in the classic sense, this would be a confrontation with sheer darkness; Plato would name it a work of "necessity" and rightly enough see it as bounding the domain of the light of knowledge with a shadow of intractable darkness. But it is not clear that it bears such tragic significance for Newton, for whom "philosophy" in this sense is not the ultimate goal of the work. That was to be the "mathematical principles of natural philosophy," and we have come to the verge: to the point at which those principles thoroughly demonstrate their own inherent limitation. Mathematics has taken its own measure. What lies beyond?

If the ultimate aim of this work is, on the other hand, to bring us closer to God, then the knowledge we seek may be of another sort, and the *Principia* may not have set out to be a work of science in the classic sense. The three-body problem has blocked our effort as theory, but if we are seeking interpretive instruments which will make it possible to read God's work in the world, we may draw a conclusion of a very different sort. The three-body problem, as Newton now shows, converts into the powerful instrument of a new order of understanding.

We have spoken at length of the role of time as an element of Newton's new mathematics: a mathematics, we said, of motion and change, for which the new method of fluxions was crucial. But time may be a dominant theme of the *Principia* in quite another way. Mathematics in its classic sense had been directed to what is timeless, to unchanging truth. Now we are on the brink of a different prospect: the truths we are after may be truths about time itself. The three-body problem is such a mathematical truth. For it demonstrates, not timeless form, but the temporal tendencies of form: with it, history becomes demonstrable knowledge. The orbits will deform, and it becomes a question for mind no longer to know the forms, as the ellipse and its elements, but the *course* of the forms, and the long effects of the perturbations.

It is this which Newton is preparing for the discussion in the third book. There, the system of the world is not a static and permanent harmony, a lasting "design," but a design which has the shape of prophetic history. The orbits are designed not for permanency but for a life of interactions, and for a lifetime which is finite and foreseeable, if not calculable, by the human mind. So, too, the planetary system is a creation which had a beginning and for which science can teach us to foretell an end. So the *Principia* takes us directly to, and itself becomes, prophetic history, and its "science" is to be understood in that context. As the mathematical principles of a science of time, the *Principia* is mathematical mirror to Scripture and prophecy. Newton is deeply committed to such interpretation and gives primacy above all to Daniel and to the Prophecy of Saint John, or the Apocalypse.[15]

---

15 Castillejo, op. cit., chap. 2. *See* Frank E. Manuel, *The Religion of Isaac Newton*, and his biography of Newton, *A Portrait of Isaac Newton*, chap. 17, "Prophecy and History." For Newton's letters to Richard Bentley on the theological implications of the *Principia* see Cohen, op. cit., pp. 271ff.

Where will the *Principia,* now read as prophetic adventure, lead us? Its consequences unfold in a series of twenty-four corollaries. Apart from the deformations in shape of the orbits of the moon about the earth, and the earth about the sun, there will be tendencies for the orbit of a satellite such as our moon to increase and decrease in inclination to the plane of the ecliptic. But then Newton goes on to imagine a ring of satellites, all subject to this same tendency. First they are conceived as fluid, a great bulge of the seas about the earth's equator, while the gravitational handle of the third body upon them becomes the effect of the tides. Next, Newton goes on to imagine that ring frozen as a solid mass. The tendency which earlier drew the moon's orbit to tilt more or less with respect to the earth's ecliptic will continue, but now it will be operating on every erstwhile moon, as component of the solid mass of this frozen sea. But this mass, in turn, which was first the moon and then the seas, may now be taken as the bulge of the earth's equator; that force which earlier worked to alter the inclination of the moon's path to the ecliptic now becomes a force exerting leverage on the equatorial bulge of the earth, which will tend to tilt the axis of the earth. The earth becomes a top, precessing under the torque of the differential gravity, first of all, of the moon and the sun.

Newton has now found the force which will account for the precession of the earth's axis, long known to astronomers, a force that expresses itself in terms of the seasons of the year as the "precession of the equinoxes"—the motion of the spring point. This can also be seen in terms of the location of the pole of the earth's axis in the heavens, or the identity of the polestar. Here is the dynamic explanation of the fact that the pole wanders over the ages among the heavens—and establishes within the framework of the *Principia* a new principle of timekeeping of the utmost importance for anyone interested in prophetic history. Here is a clock, a timekeeper set by the Creator, for the use of creatures who would one day come to know its use, ordered to the time scale of the Creation. For Newton it becomes the clock upon which the

events of the history of the world can be ordered, as well as
the clock of prophecy.

These are the new tools, then, which Newton forges out
of the wreckage of the world of timeless geometry. They will
serve him well, when he turns to the System of the World—a
system not of an ordered harmony of balanced planets, though
it bears that aspect initially, but rather more deeply and darkly
understood, a "system" of complex interconnections—in an
image favored by Newton, a "net"—committed throughout to
the working of a divine plan through the courses of prophetic
history. It bears the implications of concrete information, truth
in a form we might call "existential," less like Plato or Euclid,
and much more like bulletins of news. Newton is making us
ready for news from God.

Book II: The Refutation of Mechanism

Book II of the *Principia* is a major work of great interest in its
own right, in which Newton moves from the theory of central
forces between bodies isolated in space to that of elastic media
consisting of many bodies mutually repelling one another, or
finally—with great difficulty and limited success—of continuous,
fluid media. As in the case of Book I, many different themes
are interwoven, and many different questions are at issue. The
passage of solid bodies through elastic media, such as our air,
which Newton works out extensively under ranges of assumed
degrees and dispositions of density, has implications at one
extreme for practical artillery, but more interestingly at the
other, for the optics of refraction and diffraction. Throughout
Book II, Newton is on the track of the question of the nature
of light, which he believes must consist of particles moving in
paths modulated by interaction with some form of medium.
The greatest and overriding question of the book, however, is
that of space itself. What fills the cosmos? Is it densely filled, as
Descartes maintains, with a continuous medium—the plenum
of the mechanical philosophy? Are the planets carried by vortices in a celestial plenum (fig. 10)? Or is the cosmos cleared

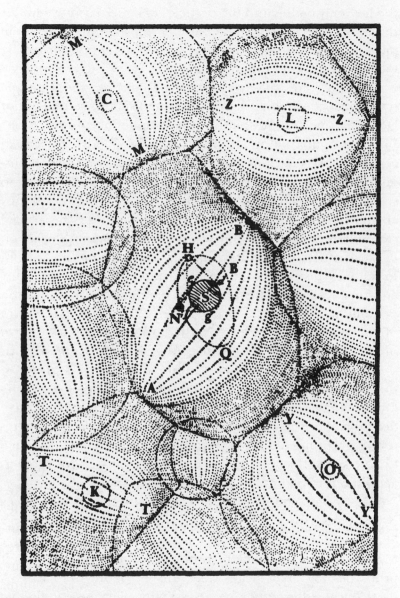

Figure 10. A figure from Descartes' *Principles of Philosophy* envisioning a world which is the dialectical alternative to that of the *Principia*. Here the cosmos is a material plenum, inviting intelligibility precisely through the absence of any further agent. Newton assigns all agency to "force," deeply related to the alchemic concept of spirit.

of all such complications and thus a great void, through which the heavenly bodies and the particles of light may move under the mystic guidance of "force" alone? Book II thus goes to the most fundamental question, and it is essential in clearing the way for the System of the World.

As I have indicated, this second book of the *Principia* has a dialectical role: it must accomplish the refutation of Descartes and the mechanical philosophy. Yet the questions I have just put are not posed merely to be dismissed; Newton takes them very seriously and carries out an extensive, painstaking series of experiments with the pendulum in order to seek out any possible effect of a pervasive ether. Though they are little noted as such, these constitute one of the crucial experiments of classical physics, the search for an interaction between ether and ordinary matter. Newton's method is first to calibrate his experiment by determining both theoretically and practically the effect of air in damping the motion of a pendulum, and then to work with pendulum bobs in the form of containers either empty, or filled with matter in one form or another. If there is a pervasive ether, it should interact with the material of the filled pendulum and reveal its presence through the retardation of the pendulum's swing. Over many cycles, a long, slow pendulum becomes an extremely sensitive detector of any such effect. Newton's negative result, therefore, constitutes in his interpretation convincing evidence that no such effect exists. Space has been cleared for force to act without impediment.

Did Newton simply wish for this negative outcome? In his own years as a student at Cambridge he had shared the world's enthusiasm at the prospect of mechanical explanations, eager to dispel vague formulations and the assumption of occult properties. By what means does the sun move the earth? A mechanical ether offers the possibility of some intelligible answer. By refuting the hypothesis, Newton is destroying the hope of intelligible explanation, and this is a hope which he cannot himself have altogether abandoned. What is at stake is the very idea of "explanation," or the question of causality. What sort of answer do we seek when we ask the question, "Why?" Is "force" (or "spirit") an answer, or only the postponement of an answer? If "force" is mathematically ordered, according to a law of force of the sort Book I has demonstrated, does that alter the status of the question? The outcome of Book II places Newton finally in the position of asserting that a mechanical explanation is out of the question, and that force, the spirit of the alchemists, is real, operative, and causal in the world. Spirit becomes intellectually acceptable in this guise, which thoroughly admits *mathesis*. The result of Book II is thus at the same time acceptable to intellect and an avenue for God's presence and activity throughout the cosmos. Book II marks the death of materialism: it has never been possible again, since Newton wrote.

We leave this then, with a fundamental issue resolved. The cosmos has been cleared of any dense ether: void, empty space must take its place. Explanation by means of any intervening matter is thereby excluded; we have seen the end of materialism. Spirit or force is essential to the operation of nature and must fill the cosmos. The connection between the sun and the planets which are bound to it is confirmed as a mystery. We may add, the operation of gravity, or any other force in nature, is at least as mysterious today as it was when Newton closed Book II and turned to write his third book, "The System of the World."

The System of the World

How strange the world of shadows and lights into whose "system" we enter with Newton! He is bringing us into the very presence of God. God's signs to us are those which the first book has taught us to read: the System is a matter of reading the messages. In the General Scholium with which the *Principia* concludes, Newton draws out these theological conclusions in a way which is entirely consistent with the work as a whole. He first sets out a set of Rules of Reasoning in Philosophy which, as we have suggested, are indeed rules for the interpretation of the phenomena as text, strikingly parallel to corresponding principles for the interpretation of Scripture. In a nature which is God's creation, the phenomena become scripture. Bringing to bear the hieroglyphs which constitute the interpreter's arsenal he has spread before us in Book I, Newton is able to see through the phenomena, penetrating to the fundamental principles to which they attest. The process is the Newtonian analysis of the phenomena—"analysis" in the sense of a motion from the obscure and the complex to the simple and best known. Once completed, it secures the foundations of a corresponding constructive argument on which a synthetic science can be built, reasoning now from first principles, which are simple and best known, to consequences which are derivative and complex. What is at once the terminus of the analysis and the beginning of the synthesis is here the law of universal gravitation: the force which joins a planet to the sun joins as well every body to every other body, however small, in proportion to their masses, throughout the world.

At the point of this watershed between analysis and synthesis, Newton asserts:

> We have discoursed above on these motions from the Phenomena. Now that we know the principles on which they depend, from those principles we deduce the motions of the heavens *a priori*. (*Principia*, pp. 420–421)

The "*a priori*" is intended, and appropriate. This is the science at which Newton has aimed throughout, concerned with "knowledge" in a new, Newtonian sense, directed to divine act, beyond mere timeless truth. It is in this mode that Newton proceeds in the System of the World to reconstruct the cosmos he has so inexorably taken apart. As in the Creation itself, the planets and their moons appear in their turns, taking their places, each its appointed motion. We demonstrate now to the concrete fact, not to the universal conclusion.

In this constructive mode, since we know all the principles, it should be possible to compute everything we might want to know. A cornucopia of practical results is indeed ready to unfold. Yet what ought to be the great and most useful triumph of this astronomical System of the World, the calculation of the future motions of the moon in the form of a reliable almanac, Newton is not able to achieve, and it is important to us here to weigh the significance of his failure. Certainly the importance of this problem to the human management of the Creation is very great. Latitudes at sea can be determined readily by taking the height of the polestar, but longitudes must be inferred from the positions of the stars in their rotations, on the basis of known times. What is needed is an accurate timekeeper which will remain reliable in the vicissitudes of travel at sea, and in Newton's time there was no such chronometer. If the moon's motions could be predicted with precision, it would serve in effect as the long-desired chronometer of universal reference; by means of times read from the positions of the moon in conjunction with an accurate table, the positions of other bodies would yield accurate longitudes. Hooke had alluded to the importance of this problem of the longitudes in his earliest correspondence with Newton, and much later, Newton himself sat on a panel to judge entries to solve this problem of "determination of longitudes at sea." Yet after immense labors, he is forced to conclude:

> But there are yet other inequalities not observed by former astronomers, by which the motions of the moon are so disturbed that to this day we have not been able to bring them under any certain rule. (*Principia*, p. 434)

How are we to weigh the significance for Newton of this lunar debacle? If truth were Socratic, the essential rewards for mind would stand in their universality, even if practical details could not be derived satisfactorily. In the domain of classical knowledge, facts are details. But Newton, operating in an altogether different arena, is entitled to no such satisfaction. If it is the role of the universal to serve the fact—if in a sense the universal is incarnate in the detail of the Creation—the economic and political outcome looms as large for Newton in proportion as the right positioning of the planets and the moons does for the Creator. Fleets were foundering for lack of Newton's numbers. Just as we said earlier that the deterioration of the perfect orbits by the action of the three-body problem did not constitute a tragedy for Newton, though it would have for the ancients, who would have regarded elliptical orbits as imperfections, now we must acknowledge the counterpart: the failure to compute accurately to the fact, that is, the failure to bring science to bear on economic, military, and political reality, is indeed tragic in a very different sense for Newton. Good governance, by God and by man, is the new criterion for a science which man now, through Newton's work, comes to share more closely with God. Newton's science is genuinely humbled by its failure, in this constructive mode, to achieve an accurate computation of the moon.

The Alchemy of the Comets

The last of the astronomical motions considered in the Newtonian construction is that of the comets. They are of special significance because, wandering widely through the heavens, they serve as probes, marvels which put all theories to the test. Curtis Wilson has recently in these pages called our attention to their importance to Newton (*GIT* 1985, 178–229). As the *Principia* draws to a close, we literally watch Newton struggle to demonstrate what has been a crucial insight in his own experience: that the comets, too, are subject to universal gravitation, that their motions are uncluttered by any remnants of a Cartesian ether, and that therefore they move in the very trajectories demonstrated in Book I (figs. 11 and 12). We see

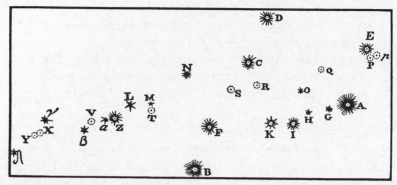

Figure 11. Newton's drawing from Book III of the progress through the heavens of the comet of 1680–81, tracked in the text on a night-by-night basis. Successive observed positions are indicated by the small circles p, P, Q, R, S, T, V, X, and Y. Newton gathered data from observers scattered over the known world, and relied heavily on the work of Edmund Halley.

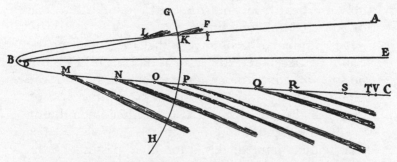

Figure 12. The trajectory Newton fits to the comet's phenomena. He carefully estimates the heat developed at perihelion in this celestial furnace. Newton shows that the comet's tail is systematically directed away from the sun, visible evidence of the distillation of spirit—the culmination at once of his alchemy and of the *Principia*. The heavens have become his laboratory.

how difficult is the practical task of fitting the orbit to observed phenomena, and how great is Newton's sense of triumph at his apparent ability to do this in the case of the comet of 1680–81, which he takes as his test.

> Hence also it is evident that the celestial spaces are void of resistance; for though the comets are carried in oblique paths, and sometimes contrary to the course of

the planets, yet they move every way with the greatest freedom, and preserve their motions for an exceeding long time, even where contrary to the course of the planets. I am out in my judgment, if they are not a sort of planets revolving in orbits returning into themselves with a continual motion. (*Principia*, p. 497)

Finally, what emerges is the sense that the comet is undergoing a form of astronomical alchemy: coming as close to the sun as Newton has computed it does, he concludes that it must have undergone immense heating, which it is nonetheless not beyond the Newtonian strategies to estimate. The comet has undergone maturation by heating at a furnace Newton could never achieve on earth:

> This comet, therefore, must have received an immense heat from the sun, and retained that heat for an exceeding long time....
> It is further to be observed, that the comet in the month of December, just after it had been heated by the sun, did emit a much longer tail, and much more splendid, than in the month of November before, when it had not yet arrived at its perihelion; and, universally, the greatest and most fulgent tails always arise from comets immediately after their passing by the neighborhood of the sun. Therefore the heat received by the comet conduces to the greatness of the tail: from this, I think I may infer, that the tail is nothing else but a very fine vapor, which the head or nucleus of the comet emits by its heat. (*Principia*, p. 522)

Newton is here recognizing that in the case of this comet, matter is being exposed to an extreme of alchemic treatment quite impossible to achieve in any earthly laboratory, yet observable in its effect, and he therefore watches the vapors which result with the greatest of interest. He concludes that he is seeing in the tail of the comet the emission of that "spirit" which was always the ultimate objective of the alchemic search and is fundamentally needed in order to complete Newton's account of the true System of the World. He conceives that the vapors from such comets run a course, ultimately condensing and falling upon the planets:

> ...comets seem to be required, that, from their exhalations and vapors condensed, the wastes of the planetary fluids spent upon vegetation and putrefaction, and converted into dry earth, may be continually supplied and made up; for all vegetables entirely derive their growths from fluids, and afterwards, in great measure, are turned into dry earth by putrefaction.... (*Principia*, p. 530)

Finally:

> ... I suspect, moreover, that it is chiefly from the comets that spirit comes, which is indeed the smallest but the most subtle and useful part of our air, and so much required to sustain the life of all things with us. (Ibid.)

Newton was not able to complete his own alchemic researches in time to balance his account of the macrocosm with a corresponding theory of the microcosm, though both belong in principle to the System of the World, but he sees by a kind of divine justice the project completed for him at a furnace in the heavens. He believes he has at last seen the spirit for which he had for so long been searching, and which bears the most important force that must be included in any full account of nature, a repelling or elastic force. He returns to the question of this spirit in the closing words of the *Principia*, as we shall see.

The General Scholium

The *Principia* closes with a frankly theological discussion, in what Newton terms the "General Scholium"—that is, the scholium to the work as a whole. This may be regarded, by

those who take the *Principia* to be the first work of modern
physics, as a mere commentary or afterthought, an interesting
but unnecessary addendum. I think rather that for Newton it
is central to the purpose and method of the entire work. It
is brief and tightly written, and the reader is invited to turn
directly to it. Newton first recapitulates the argument against
vortices, with the result that:

> ...all bodies will move with the greatest freedom;
> and the planets and comets will constantly pursue
> their revolutions in orbits given in kind and position,
> according to the laws above explained; but though
> these bodies may, indeed, continue in their orbits by
> the mere laws of gravity, yet they could by no means
> have at first derived the regular position of the orbits
> themselves from those laws.
> ...it is not to be conceived that mere mechanical
> causes could give birth to so many regular motions.
> (*Principia*, pp. 543–544)

In discussing the concept of law earlier, we concluded
that the structure of the *Principia* belongs to the spirit of the
Judeo-Christian tradition. Newton now, holding this mirror
to his work, reveals the sense in which that is true. Newton
has shown the cosmos to be cleared of mechanisms, and thus
liberated for the agency of spirit. This agency he now calls
"dominion":

> And from his true dominion it follows that the true
> God is a living, intelligent, and powerful Being.... He
> is eternal and infinite, omnipotent and omniscient;
> that is, his duration reaches from eternity to eternity;
> his presence from infinity to infinity; he governs all
> things, and knows all things that are or can be done.
> ... He endures forever, and is everywhere present;
> and, by existing always and everywhere, he constitutes
> duration and space. (*Principia*, p. 545)

Readers must balance for themselves the elements of Newton's account. I am impressed on this reading with Newton's awe and the sense he conveys of the immediacy of mystery. What he says of his God hardly goes beyond what he has seen and demonstrated—but he has seen and demonstrated so much:

> We have ideas of his attributes, but what the real substance of anything is we know not. In bodies, we see only their figures and colors, we hear only the sounds, we touch only their outward surfaces, we smell only the smells, and taste the savors; but their inward substances are not to be known either by our senses, or by any reflex act of our minds: much less, then, have we any idea of the substance of God. We know him only by his most wise and excellent contrivances of things, and final causes; we admire him for his perfections; but we reverence and adore him on account of his dominion: for we adore him as his servants; and a god without dominion, providence, and final causes, is nothing else but Fate and Nature. (*Principia*, p. 546)

Finally:

> And thus much concerning God; to discourse of whom from the appearances of things, does certainly belong to natural philosophy. (Ibid.)

It would be an error to think that this is all that Newton has to say about God, and that he has in view only what we might call a "natural religion." We know that he thought and wrote extensively on questions of Christian faith and the interpretation of Scripture, about which he held strong if idiosyncratic views. The *Principia* must be seen as one aspect of this faith, essentially consistent with it. The two complement one another, two channels of Newton's relation to his God. Probably a stronger statement is justified: Newton could not have arrived at the *Principia* without a foundation in the faith he brings to it from theology. His concept of "law" derives ultimately from his sense of the God of Scripture, while his faith in the simplicity of nature and the confidence in the phenomena, on which his science is founded, rest finally on his faith in God. Newton has spread before us a Divine Comedy; without God, such a comedy would have no foundation. To "discourse" of God "from the appearances of things" is appropriate for Newton. For Newton, that study itself is only possible as God's gift.

The General Scholium, and with it, the *Principia,* close not on this overtly theological reflection but with two questions which may seem of a very different sort, though in truth they are not. The first concerns the cause of gravity. To this, Newton says, he has no answer:

> And to us it is enough that gravity does really exist, and act according to the laws which we have explained, and abundantly serves to account for all the motions of the celestial bodies, and of our sea. (*Principia,* p. 547)

Some may feel that this whole paragraph indicates that Newton is still seeking a mechanical explanation, and it is true that he works in other places with various notions of ethers in considering the problem of attraction. I think, however, that the sentence above is decisive: he is ready to rest content in the presence of mystery, and though at this point he has only made a beginning, he does not envision a completed natural philosophy which terminates on a note essentially different from the one he has sounded in this General Scholium. Mystery does not represent a failure of natural philosophy but its rightful culmination.

The final question is the most intriguing, and like the closing observation concerning the comet, takes us back to the core of his alchemic conviction. If our present reading of the *Principia* is correct, it is highly appropriate that it is on this alchemic theme that the work closes. Newton writes:

> And now we might add something concerning a certain most subtle spirit which pervades and lies hid in all gross bodies.... (*Principia,* p. 547)

As his description of this "spirit" unfolds, we see how his thinking is running. It is to account for a very wide range of phenomena, including on the one hand the optics of diffraction and refraction, and on the other, animal spirits, the action of sense and will. Newton is inviting something like the electromagnetic ether, but something more as well, indeed the very principle of life itself. How, might we imagine, would this comport with the rest of the System of the World, if he had been able to include it?

Though it would be associated with and borne by subtle particles, we may safely assume that Newton is not speaking of a material ether; when he says "spirit," he means just that. In the alchemic context, such spirit, extracted from common air, serves to vivify inert materials and is essential to the processes of life. Spirit as an immaterial, nonmechanical vital principle is fully compatible with all we have learned from the rest of the *Principia,* and with the convictions of the Newtonian alchemy on which it rests. Newton's mathematics of nascent quantities, living and growing, is by no means coextensive with mechanism and blind determinism; we have tended to forget the possibility that mathematics may speak equally of life and mind.

It is appropriate that Newton withholds further discourse concerning this spirit, this second element of the system of his world, to which he was never able to break through. There is a note of awe in the hush which falls over the *Principia* in these closing words; God has more yet to tell mankind. The work of revelation, the prophetic tradition to which I think it is clear Newton believes the *Principia* belongs, has a further course to run. Looking ahead in this last passage, as in the final "Queries" he appended to the *Opticks,* Newton seems to bring us to join him at the threshold of the temple, where it is necessary and right that silence reign.

Maxwell

Introduction to the Maxwell Essay

I N THE ESSAY ON NEWTON I described a thinker for whom "nature" denoted something that was, at all levels, suffused with *spirit*. Nevertheless, as I have already indicated in the Preface to these essays, such was not at all the Newton young James Clerk Maxwell was meeting as a student at Cambridge in the mid-1800's. With all talk of spirit long since forgotten, Newton's laws and methods had become the working medium of a rigorous mathematical discipline to which aspirants to a university degree were subjected as a form of mental training, largely apart from any further interest in natural philosophy. Success in the culminating tripos examinations, on which career prospects in any field might depend, was a matter very largely of quick responses to standard problems.

Within serious science itself in England, and to a large extent on the Continent as well, the Newtonian paradigm reigned supreme: success in the solution of a new problem in the sciences was felt to depend on getting it into Newtonian form—specifically, to characterize it in terms of a force acting between two bodies, the magnitude of the force being expressed as a function of the distance between them. It was by far best if that function proved to be the inverse square law, as in the paradigmatic case of gravity. Indeed that very success had been spectacularly achieved through the discovery of the laws of the force between magnetic poles and the force between electric charges. This, then, was the world according to Newton into which Maxwell was being inducted. For various reasons, it did not particularly please him.

It did not please him, for example, that the approach came in the form of dogma; for Maxwell was very much interested in questioning the bases of things. He had arrived at Cambridge after two years in Edinburgh, where he had immersed himself with earnest enthusiasm in Scottish metaphysical skepticism, a taste which never left him. Neither did it please him, as a Galloway Scotsman, that the approach was associated with a highly aristocratic and socially selective British university system.

Most fundamental, however, was surely the sterile form that Newtonian theory had taken, reduced to hardly more than a stark mathematical formalism. With Newton's *spirit* gone and nothing to take its place, a Newtonian force was now taken to represent unmediated *action-at-a-distance*, something Newton himself had declared no one of sound mind could accept.

In the face of that demand for uncritical acceptance of formulaic *action-at-a-distance*, the combination of Maxwell's aversion to social aristocracy and his dislike of pedagogical tyranny moved him, I believe, to strong rebellion. True, he skillfully embraced the mathematical powers of the Newtonian methods when they served his purposes, and he welcomed every insight into the workings of nature; but Maxwell's relation to the Newton he knew may have resembled that of the historical Newton to Descartes: admiration accompanied by a strong and quite specific aversion—a dialectical *negation*, I will argue—out of which, in Maxwell's case, field theory was to be born.

In forming that revolutionary theory, Maxwell gave a central role to the methods of Michael Faraday, who was not only a commoner nearly devoid of formal education, but by the world's standards (and his own admission) altogether *unmathematical*. Thus Maxwell's rebellion was theoretical, pedagogical and social, all at once. But that is a story the following essay will tell.

Maxwell's *Treatise* and the Restoration of the Cosmos

J AMES CLERK MAXWELL'S *Treatise on Electricity and Magnetism* is clearly a great book—and, at the same time, clearly a great problem. Ever since it was published, in 1873, it has stood as something of an enigma to even the best-intentioned of readers. Although there seems to be no question that the *Treatise* is a classic of physical thought and marks a turning point in our understanding of the natural world, the editors of the *Great Books of the Western World* chose not to include it in their collection; they were no doubt quite right in this judgment in terms of the book's likely usefulness to their readers. Maxwell wrote with constant concern for questions that are no longer regarded as valid, and in this text on electricity and magnetism he sought to achieve philosophical goals which modern readers do not share, and which only seem to render his account obscure. There is even room for doubt whether it is a "book" at all, or some other kind of published entity, a product of the tempo and pressures of the life of science in the modern world.

Thus it is with some misgivings, mixed with a kind of reckless courage, that I invite readers of *The Great Ideas Today* to join me in my journey to a possible-book which I believe is indeed a great work of literature and no less beautiful for the perplexities it incorporates. I have thought that the *Treatise* is a little like a city—not the ordered *polis* which is Plato's Republic, but a teeming modern city, crowded with spirit and life, with activities of all sorts, some converging upon one another, others starting out in directions of their own with remote and unknown destinations, and all guided by a variety of leaders of the present and mythic heroes of the past. If the *Treatise* is such a city, it is an image of what we call "modern science," and if a darkness hangs over Maxwell's work, it may be the shroud that has shadowed our belief in science as a prospect for mankind.

It is popular these days to speak of "scientific revolutions" and their structures, and the renewed recognition of these radical turns of thought has certainly very much illuminated our study of the history of science. Such a "revolution," such a radical turn of thought, is indeed taking place in the pages of Maxwell's *Treatise,* and the fact that the foundations of thought are in flux in the very course of the work itself is part of both the interest and the difficulty of reading it.

But is there not something too pallid in this talk of changes of paradigm? For they are not simply happening, as response to crises in the interpretation of phenomena; the best of them are deeply motivated dialectical thrusts of the human spirit, and the scientist makes his groping way forward under the spell of human concerns much larger than the needs of any special science, or even of science itself. At least, I feel that is the case with the *Treatise.* It is a very human work—unabashedly so, with a major place given, especially, to the warm, genuine figure of Michael Faraday. In its humanity, the *Treatise* seems to me to reach out to some of the largest human questions. The revolution taking place in the pages of this work is turning the world inside out, challenging and perhaps reversing the most fundamental understanding of the nature of science itself, and—I think—undertaking to restore something of the fullness of the cosmos which the Newtonian revolution had, in a sense, emptied with its talk of forces and action-at-a-distance.

With the establishment of the concept of the *field,* we might see the end of the Age of Newton and the beginning of the Age of Space, regarded not as void but as configured terrain admitting curvature of mass and density.

But things are not so simple. Entering upon this new world evidently entails totally revising our conceptual structure: we can get the old cosmos back again only in wholly new terms. Here Maxwell's special indirectness and a certain deeply rooted skepticism of the claims of conventional wisdom serve him well in opening the way to a new mode of thought. I see this as a metaphorical transformation of the old, Newtonian physics, so that, remarkably enough, there is included within the pages of Maxwell's *Treatise* a kind of "New *Principia*," of which the application to electromagnetism in the *Treatise* itself is only one illustration.

Maxwell was a deep admirer of Newton ("that mind without a blemish"), and in certain ways the structure of the *Treatise* becomes assimilated to that of the *Principia.* But Maxwell's "New *Principia*" completely reverses the old, and it does so through such a deliberate and conscious use of analogy and metaphor that it is as if the old *Principia* had come under a spell and were being transformed in a comic Forest of Arden or on Prospero's island. It is out of such a magic transformation that Maxwell draws his revolution, from Newtonian action-at-a-distance and a physics of intangible law, to a restored cosmos once again made whole by contiguity and communication throughout. Ultimately, I think, this implies for Maxwell a new order of intelligibility, not only in the sciences, but for the understanding of history and the human community.

In thus identifying theories of action-at-a-distance as "Newtonian," I do not mean to be making a statement about the dialectical richness of Newton's own thought, as revealed for example in the General Scholium to the *Principia,* or certain of the "Queries" appended to the *Opticks.* I am thinking rather of the Newton known to the world through the *Principia* as the book of distance-action determined by strictly formulated mathematical law, and of the long, impressive scientific tradition founded upon this perception. Newton himself certainly speculated relentlessly about possible media, aethereal or otherwise, which might operate between a center of force and a body attracted to it, as between the sun and a planet. But we should remember that Newton had no desire to reduce the operations of nature to mechanics; in particular, the Second Book of the *Principia* is written as a refutation of Descartes and his vortices. When Newton asks,

> What is there in places almost empty of matter, and whence is it that the Sun and planets gravitate towards one another, without dense matter between them? (*Opticks,* Query 28)

the ultimate answer he is seeking is not anything physical, but "a Being incorporeal, living, intelligent, omnipresent...." Newton is on the track of God. The question of Newton's own understanding is complex, and we cannot pursue it here, but I would suggest that for Newton, physics rightly ends with formal law and action-at-a-distance precisely because this delivers us into the immanent dominion of a being who is "God of Gods, and Lord of Lords" (*Principia,* General Scholium, p. 544).

In part, this essay will be an inquiry into the question of whether mathematical physics has a "rhetoric" and will submit to a kind of literary criticism. The "rhetoric" I have in view does not concern ornament, does not exist as an *addition* to the scientific statement, but is intrinsic to it, as the very means by which the statement is made. One point of view would have it that in science the "matter" alone counts, and that questions of "manner" are distractions. This would be a way of understanding the positivist program and would fall in with Hertz's remark, made no doubt from desperation in his attempts to understand Maxwell on the subject of electric charge: "Maxwell's theory is Maxwell's system of equations." Pierre Duhem wrote an entire book, a polemic against Maxwell, in which he attempted to purge electromagnetism of the rhetorical mists in which Maxwell had, in his view, wrapped his subject.[1]

I am encouraged in the alternative view by Maxwell himself, who frequently revealed his preoccupation with questions of form and style. Such questions arise whenever the same thing can be said in more than one way and the choice among these ways makes a difference. Consider this statement from Maxwell's address to the British Association:

> As mathematicians, we perform certain mental operations on the symbols of number or of quantity, and, by proceeding step by step from more simple to more complex operations, we are enabled to express the same thing in many different forms. The equivalence of these different forms, though a necessary consequence of self-evident axioms, is not always, to our minds, self-evident; but the mathematician ... can often transform a perplexing expression into another which explains its meaning in more intelligible language. (*SP* ii/216–17)[2]

Maxwell is concerned to master the symbols themselves—to learn to use "analogy," "illustration," and mathematical transformations to wring meaning out of the seemingly abstract and empty forms of mathematical physics, and thus to preserve science for philosophy and the human understanding.

The *Treatise* is, I believe, the work in which he brings to bear all the arts he has developed for this purpose. It is frankly a work dedicated to a rhetorical purpose: Maxwell knows,

---

[1] Pierre Duhem, *Les théories électriques de J. Clerk Maxwell.*

[2] This short-form notation will be used throughout to refer to W. D. Niven, ed., *The Scientific Papers of James Clerk Maxwell.*

and explains in his Preface, that there are two general ways, *mathematically equivalent,* in which electromagnetic theory can be presented. They predict the same phenomena; Maxwell acknowledges at the outset that even the most distinctive aspect of his work, the electromagnetic explanation of light, may be predicted equally by the retarded potentials of Ludvig Lorenz. But the difference between two such "equivalent" accounts is precisely the source of the personal energy with which Maxwell writes his *Treatise*; his purpose is to reveal the intellectual consequences of following just one of these alternatives consistently:

> These physical hypotheses, however, are entirely alien from the way of looking at things which I adopt, and one object which I have in view is that some of those who wish to study electricity may, by reading this treatise, come to see that there is another way of treating the subject…. (*Tr* i/x)[3]

Throughout the *Treatise,* Maxwell throws his equations into that form which will "suggest the ideas" he wishes to present to the reader's mind. These are Faraday's ideas, and the *Treatise,* in my view, can best be understood as a strategic effort to transform the equations of electromagnetic theory into a new structure that captures Faraday's geometrical and physical vision.

Maxwell distinguishes his purpose from that of one who treats all mathematically equivalent forms as equals by saying in the Preface:

---

[3] Short-form notation for reference to Maxwell's *Treatise on Electricity and Magnetism.* The first edition was published in 1873. The edition referred to is not the first edition, however, but the third, edited by J. J. Thomson and published in 1892. It is this edition which has been photoprinted and reissued by Dover Publications, and which is currently in print.

> I have therefore taken the part of an advocate
> rather than that of a judge, and have rather exempli-
> fied one method than attempted to give an impartial
> description of both. (*Tr* i/xi)

It is Maxwell's *advocacy* that I think constitutes this a truly
dialectical work. It is governed finally, not by questions of
mathematics or laboratory evidence, but by the negation of
one path and the affirmation of another according to human
criteria not found in the symbols or in the data.

I hasten to remark that this is just one aspect of the *Treatise*,
and that in reading it in this rather abstract way as a work of
thought we are setting aside the sense in which it very much
springs from, and feeds back into, the England of capitalism
and the Industrial Revolution. Perhaps in an effort to find a
"dialectical" standpoint, we are eliminating from consideration some of the very clues that would lead us to a sense of the
larger dialectic in its wholeness, in which the ultimate identity
of Maxwell's revolution and the Industrial Revolution would
appear. Another reading of the *Treatise*, which I hope to do on
another occasion from the point of view of its intersection with
measurements, systems of units, economics, and technologies,
would give a different picture, complementary to this one. The
disarray that readers have found in the book is in part the consequence of these seemingly disparate, but ultimately deeply
related, factors. But not everything can be done at once, and
we will do well if we can read the *Treatise* at this point from even
one consistent point of view.

Faraday and the Goal of the *Treatise*

In an extraordinary way, Maxwell's *Treatise on Electricity and
Magnetism* presupposes that the reader is already familiar with
another book, Faraday's *Experimental Researches in Electricity*.
Many works refer repeatedly to a single source, but there

must be few that *command* their readers as Maxwell does in a footnote: "Read Faraday's *Experimental Researches*, Series i and ii"—advice that might very well apply to any reader of the present essay.[4]

For Faraday's work stands as the definition of the goal of the *Treatise*: in some essential way, Maxwell must capture, and as nearly as possible complete, Faraday's project. Thus, the *Treatise* can be seen as a translation into more formal, mathematical terms of Faraday's experimental narrative. But I do not think, as I once did, that this in itself would adequately characterize the relation between the two works, or even that Maxwell would think that translation into formal mathematics or ordered theory would in themselves constitute an improvement on Faraday's original. Rather, it seems to me that Maxwell sets out to capture the spirit of Faraday's work, and to carry it through, as nearly as possible, to the completion Faraday himself sought. In so doing, Maxwell utilizes extensive mathematics and creates something like a formal theoretical system, but these, I think, are merely means to bring greater power to bear on the solution of Faraday's own problem, which, as we shall see, is in no way either analytic or theoretical. Maxwell expresses his dedication to Faraday in strong terms:

> If by anything I have here written I may assist any student in understanding Faraday's modes of thought and expression, I shall regard it as the accomplishment of one of my principal aims—to communicate to others the same delight which I have found myself in reading Faraday's *Researches*. (*Tr* i/xi)

[4] Short-form notation for reference to Faraday's *Experimental Researches* will be *XR*, followed by paragraph number. Any reader would do well to read not only Series I and II, as Maxwell recommends, but Series XI as well, and of course to follow the lead of the references made in the course of this article.

It is therefore essential that we begin with some attention to Faraday and his *Researches,* in order to place before ourselves the goal that Maxwell proposes. We have, in effect, to begin with the wrong book: with the *Researches* instead of the *Treatise.*

Faraday developed two concepts which became increasingly significant for him. One was of course his concept of "lines of force." He was impressed by this notion early when he spoke of the lines merely as representations. Toward the end of his work, they bore nearly the whole burden of his thought, and he speculated increasingly about their possible "physical existence." The other fundamental concept was that of the "electrotonic state." This is now unfamiliar to students of science, but for Faraday it was both significant and deeply troubling, and he returned to it, as Maxwell points out, again and again from its introduction in connection with the induction experiments of Series I in 1831, to the end.

In these early experiments, Faraday had been the first to discover that if two circuits are adjacent, the first carrying current and the second including a galvanometer, when the current is interrupted in the first, a brief current is *induced* in the second, as indicated by a deflection of the galvanometer. Essentially, Faraday was convinced that the surge of current in the secondary could not occur unless the primary current had held the secondary circuit in a state of electric *tension,* which was relaxed when the primary current ceased. It was the relaxation of that state of tension which, like the release of a tense spring, caused the galvanometer needle to jump. Faraday had a way of sending off to the university to have a proper scientific term devised to fit his need and, in this case, the result he got was *electrotonic,* from the Greek *tonos,* tension, as in the string of a drawn bow. As we shall see, Maxwell's translation of Faraday's thought turns about these two ideas, the "lines of force" and the "electrotonic state." Maxwell thinks he has found the physical

state, the electrotonic state, Faraday was looking for—though
not quite in the form Faraday had expected.

For Faraday, these concepts do not function as the elements
of a connected *theory* in any formal sense. Faraday was almost
totally uneducated in mathematics, a fact that fairly directly
reflects the circumstance that he belonged by birth to the working class. It is difficult to grasp the significance of this fact—at
once intellectual and social—for science. However brilliantly
he succeeded in self-education in other areas, Faraday apparently never felt it necessary to acquire the mathematics he had
missed. The result is that he does not have before him with
any vividness that universal paradigm of a reasoned deductive
system, the geometry of Euclid. He has no working notion of a
system of axioms and postulates, or of reasoning leading with
logical rigor to universal theorems.

It seems to me that there is in this something of a deliberate rejection, in a spirit which Maxwell may have shared
and which may have helped to draw Maxwell to Faraday.
We are blessed with the record of a remarkable, revealing
exchange of correspondence between Faraday and André-
Marie Ampère, the French philosopher and mathematical
physicist whose work on electrodynamics Duhem once called
"a theory which dispenses with the Frenchman's need to envy
the Englishman's pride in the glory of Newton." Here Faraday
confronts his opposite, a brilliant mathematical theorist, and
in the course of the correspondence Faraday describes his
view of his own role:

> … I regret that my deficiency in mathematical knowl-
> edge makes me dull in comprehending these subjects.
> I am *naturally skeptical in the matter of theories* and there-
> fore you must not be angry with me for not admitting
> the one you have advanced immediately.[5]

---

[5] Faraday to Ampère, Feb. 2, 1822. L. Pearce Williams, ed., *The
Selected Correspondence of Michael Faraday*, 1, p. 132. Italics added.

This skepticism of theory, I believe, turns him away from mathematics almost on principle. Faraday tends to portray mathematicians as operating on a height, while his own work, as experimentalist, lies below, close to nature and to fact. It would be a mistake, I believe, to overlook the element of pride that mixes with humility in his descriptions of his supposedly more modest work. In a moment of triumph following upon the discovery of electromagnetic induction, he wrote to a friend:

> ... It is quite comfortable to me to find that experiment need not quail before mathematics, but is quite competent to rival it in discovery; and I am amazed to find that what the high mathematicians have announced as the *essential condition* [of one of the phenomena of induction] ... has so little foundation.... [6]

There is more than a suggestion of a moral note in this rejection of theory; in a phrase he used without any hint of apology, Faraday was an "unmathematical philosopher." He speaks an altogether different language from that of the theoretical scientist. He is not stepping hesitantly toward a mathematical physics; he is marching confidently along a different road.

In retrospect, this is a fantastic situation. In the land of Newton, at a time when mathematical physics was again flourishing and the powerful techniques of analytic mathematics had at last made their way from the Continent to England, some of the most creative scientific work of the century was done over a long period of years by a man who had no notion of the *Principia* and did not share either Newton's goals or any appreciation of the power of analytic methods. For Faraday, Newton's classic triumph was essentially meaningless, that

---

[6] Henry Bence Jones, *The Life and Letters of Faraday*, 2, p. 10.

triumph which had polarized the intellectual life of Europe for a century. And Maxwell, who had taken high honors in mathematics at Cambridge, chose this unmathematical philosopher as guide for his own definitive treatise!

How can we understand such a dramatic break in any orderly continuity in the development of science? I think we are witnessing a dialectical questioning of the concept of science itself. In identifying with Faraday's project, Maxwell is not only articulating a new theory of electromagnetism; he is giving science a deliberate new shape.

What, then, is Faraday's concept of science? Despite his proclaimed role as "experimentalist," it is clear that he had one of the most fertile and insistent of speculative minds; in a certain sense, he was constantly producing new hypotheses and his mind was constantly reasoning from them. The result of this, reported in the thousands of paragraphs of the *Experimental Researches* and the equally numerous pages of the *Diary,* is not theory but a vast weaving and unweaving of suspected powers, a process of continual discovery and identification, a great, highly unified formulary for the production and classification of effects.

Faraday, as his biographer John Tyndall proclaimed and all the world agreed, is the great "discoverer"; the paradigm for Faraday is Odysseus rather than Euclid: in a sense he, too, travels from land to land, reporting wonders, guided by legend and myth, rumor or divine love. For Odysseus, the dominant desire is to see men's cities and to know their minds, and to gather all this together in the return to Ithaca. For Faraday, it is to investigate all the powers of nature, and to unveil them as essentially one in the lecture hall at his Royal Institution on Albemarle Street. This is not theoretical physics; as has been suggested, it is essentially chemistry, not in the modern sense of Lavoisier, but as a science of powers in the tradition of van Helmont and Stahl.[7]

---

[7] This point is made by L. Pearce Williams in Lancelot Whyte, ed., *Roger Joseph Boscovich*, p. 163.

Faraday recounts his adventures not, I think, as theorist but as *interpreter* of nature, relying on the inherent intelligibility of good narrative. This is science as interpretation, not as theory: what we might call, drawing upon Heidegger, *hermeneutic* science.

The *Experimental Researches* is dense with questions—Faraday's method is that of unremitting inquiry. The very notion of a series of researches is, in a sense, that of a chain of linked questions and answers. The underlying question for Faraday is ultimately always the same: what really exists in nature? The practical form that this takes is the *test*: what will happen if I do *this*? Can I produce the phenomenon, the visible or tangible evidence, that will be the sure symptom of the existence of this or that suspected power or state? When the plane of polarized light was rotated on passage through Faraday's "heavy glass" in a strong magnetic field, he announced that he had:

> ... at last succeeded in *magnetizing and electrifying a ray of light, and in illuminating a magnetic line of force.* (*XR* ¶2148)

This I believe is symbolic of Faraday's concept of science: to make manifest to the eye what is suspected to exist in nature:

> For if there be such physical lines of magnetic force as correspond (in having a real existence) to the rays of light, it does not seem so very impossible for experiment to touch *them*; and it must be very important to obtain an answer to the inquiry respecting their existence, especially as the answer is likely enough to he in the affirmative. (*XR* ¶3305)

A hypothesis or theory is nothing more than an unresolved suspicion, a part, as Tyndall suggested, of the scaffolding, not of the edifice of science itself, which deals with existences.

Throughout the *Researches,* Faraday sought what he called "contiguity" in nature; understandably, he seeks the same contiguity in the *account* of nature. A work of science should record a completed exploration, a detailed mapping, without gaps, of contiguous substances and powers. It is not merely Faraday's clarity of view and inventiveness that attract Maxwell, but this image of the form physics might take as a physics of contiguity: in a sense, a "field" model of physical explanation itself. In the transformation Maxwell achieved of Faraday's thought, I believe his objective was not only to find mathematical expressions appropriate to Faraday's concepts but to give analytic form to Faraday's vision of an a-theoretic physics through a completed expression of the idea of the electromagnetic field (Figure 1).

Of the two concepts that are so important in Faraday's thought, the "lines of force" and the "electrotonic state," it was curiously the less known "electrotonic state" that seemed to Faraday himself the more fundamental idea, and it is this latter, as we shall see, that Maxwell takes as the key to the translation of Faraday. As we have remarked, it appears early in the *Researches* as the expression of Faraday's perplexity at finding that it is a *change* in the primary current that causes induction. Like other experimenters, he had earlier sought an effect from a *steady* primary current; when, instead, it was the interruption of the primary current that produced the effect, he continued to perceive the pulse in the secondary as evidence of the change in this expected, though never observed, "state." He cannot get this "electrotonic state" out of his head:

> Whilst the wire is subject to either volta-electric or magneto-electric induction, it appears to be in a peculiar state.... This electrical condition of matter has not hitherto been recognized, but it probably exerts a very important influence in many if not most of the phenomena produced by currents of electricity. (*XR* ¶60)

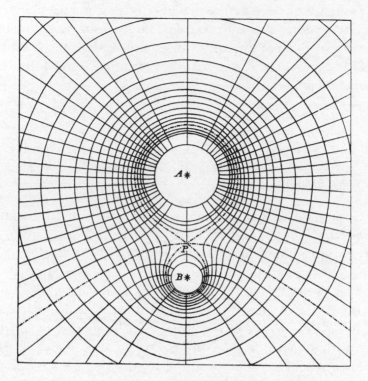

Lines of Force and Equipotential Surfaces.

A = 20   *B* = 5.   *P, Point of Equilibrium.*   *AP* = ⅔*AB*.

Figure 1. The first figure of James Clerk Maxwell's *Treatise on Electricity and Magnetism*.

He has to confess, however, that he found no evidence whatever for the existence of this newly named and announced state:

> This peculiar condition shows no known electrical effects whilst it continues; nor have I yet been able to discover any peculiar powers exerted, or properties possessed, by matter whilst retained in this state. (*XR* ¶61)

He never did. In Series II, he formally withdrew the claim of its existence, though before making his reluctant retraction he had made great efforts to reveal it: he feels strongly that he is in touch with a reality that is present even when the magnetic field is completely at rest. Thirty years after Series I, he reasserts this faith:

> Again and again the idea of an *electro-tonic* state ... has been forced on my mind; such a state would coincide and become identified with that which would then constitute the physical lines of magnetic force.... (*XR* ¶3269)

Should we say that the "electrotonic state" was, for Faraday, a *theoretical concept* and conclude that Faraday was, after all, a theoretician despite himself? Rather, I think this "state" is theoretical for Faraday only insofar as there is a gap that has not yet been filled in the explanatory series. There is "a link in the chain of effects, a wheel in the physical mechanism of the action, as yet unrecognized." The goal is to fill that gap with something *other* than a speculative concept. As a science reaches completion, theory disappears. Completed science is a-theoretical. With this, I think Maxwell is in deep agreement.

It is useful, finally, to notice what Faraday does *not* ask himself. He does not ask questions about quantitative, functional relationships—about "laws." He does not work with ratios and proportions. Not only does he almost never write an equation, he never asks the kind of question that has an equation as the natural form of its answer.

Faraday's discomfort with the notion of a functional relation in mathematics is revealed poignantly by a remark he made very late in his career, at a time when he had finally been brought into confrontation with the, to him, dreaded and offensive inverse-square law of gravity. He rebels at the formulation: "with a strength VARYING INVERSELY...." The capital letters are his, expressing his outrage at what he considers a blatant violation of the principle of *conservation* of the force: how can it then "vary"? He understands, indeed, the algebraic relation as describing the effect, but the proportion, which for Newton and many generations of scientists after him had fully characterized the force, seems to Faraday utterly unjust to it. "Why, then, talk about the inverse square of the distance?" he says, commenting on a dismissal of his own account by the astronomer-royal, Sir George Airy. "I had to warn my audience against the sound of this law and its supposed opposition on my Friday evening...."

This is of course a naive view, profoundly naive. Faraday had said:

> Let the imagination go, guiding it by judgment and principle, but holding it in and directing it by *experiment.*[8]

This remark, in its context in the *Diary*, is not a methodological reflection but a Dionysian outcry in the midst of the chase. It is surrounded by a cascade of ideas, as much visionary as experimental, about a wished relation between gravity and electricity. Faraday built the world of the *Experimental Researches* on this assurance of the reality, immediacy, and unity of the world of nature. It is, perhaps, the literary triumph of Maxwell's *Treatise* that he is able to carry this vivid, human insight into mathematical physics itself with such gentleness and conviction, in the process creating a new vision at once of mathematical physics and our relation to nature—even, perhaps, to each other.

A Mathematical Figure of Speech

To "fix our ideas," in a phrase of Maxwell's, we would do well to consider a definite experimental arrangement, as in

---

[8] Thomas Martin, ed., *Faraday's Diary*, 7, p. 337.

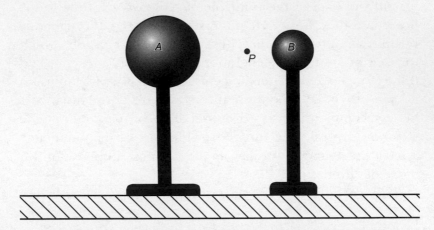

Figure 2. Two positively charged metal spheres, mounted on insulating stands: A = +20; B = +5. Turn this drawing on its side and you have the arrangement of Figure 1. The figures are printed to scale, so that Figure 2 superimposes on Figure 1.

Figure 2. Here the circles labeled "*A*" and "*B*" represent two positively charged metal spheres, *A* with a charge of 20 units, and *B* with a charge of 5. There is a force of repulsion between them: each is in some way or other able to exert an influence at a distance. For example, *A* could be made to induce a charge on some neutral body at a distance, and Faraday thus tends to refer to this distance-action in terms of "induction." How can we understand the action of a charged body over an intervening distance?

For Faraday, the right interpretation of this situation is crucial. He cannot believe that it is adequately explained by the invocation of a mere mathematical law of action-at-a-distance; something must intervene:

> … I was led to suspect that common induction itself was in all cases an *action of contiguous particles* and that electrical action at a distance … never occurred except through the influence of the intervening matter…. if this be true, the distinction and establishment of such a truth must be of the greatest consequence to

our further progress in the investigation of the nature of electric forces. (*XR* ¶¶1164–1165)

To carry through the interpretation of electrostatics in Faraday's terms, as the action of an intervening medium, is the challenge Maxwell takes up in Part I of the *Treatise*. Maxwell explains that he will start from the position of traditional theory:

> In the following treatise I propose first to explain the ordinary theory of electrical action, which considers it as depending only on the electrified bodies and on their relative position, without taking account of any phenomena which may take place in the intervening media. In this way we shall establish the law of the inverse square, the theory of the potential, and the equations of Laplace and Poisson. (*Tr* i/62)

This we may refer to as the Old Way, and think of as the way of Newton and the Newtonians, who included most of the mathematicians of electricity in Maxwell's time. It is important to reflect, as mentioned earlier, that Newton's own view of force laws and distance-action was not simple; Maxwell was very much interested in the fact that Newton seems at times to have covered some of the ground that Maxwell himself was about to explore. Here, I am thinking primarily of the tradition of mathematical physics built on the foundation of the *Principia* and formal laws of force acting at a distance.[9]
Newton had concluded in the *Principia*:

> That there is a power of gravity pertaining to all bodies, proportional to the several quantities of matter which they contain (*Principia*, Book III, Prop. VII, p. 414)

and that

---

[9] For a study of the development of Newton's actual thinking in these matters, see the valuable article by Curtis Wilson in last year's edition of *The Great Ideas Today*, "Newton's Path to the *Principia*" (*GIT* 1985, pp. 179–229).

> In two spheres gravitating each towards the other, if the matter in places on all sides round about and equidistant from the centres is similar, the weight of either sphere towards the other will be inversely as the square of the distance between their centres. (*Principia*, Book III, Prop. VIII, p. 415)

This is Newton's law of gravitation. The corresponding law governing the action of one charged body upon another was given by Coulomb. It had been the great triumph of the early heroes of the mathematical theory of electricity to be able to state such a law for charged bodies and to use it to account for observed electrical phenomena, thereby bringing electrostatics within the fold of the only known mathematical physics, that of an action between two centers of force, along the line which joins them. If we insist on asking *how A* influences *B*, we get only the formal, blank answer, "by virtue of a law...." Indeed, in the *Treatise* Maxwell, true to his promise, develops the theory of such electrical laws very thoroughly. But his intentions lie elsewhere.

To move beyond Newton, Maxwell proposes to use what we might think of as a mathematical figure of speech. He will use a certain mathematical theorem as a bridge between the Old Way and the New, namely that relationship he calls "Thomson's theorem." Without entering here into the details of the mathematical odyssey on which Maxwell embarks, we might simply note that he draws this theorem from a certain storehouse of treasures, a fascinating section, labeled simply "Preliminary," which precedes chapter 1 of the *Treatise* and seems to parallel the Lemmas at the outset of Newton's *Principia*.

The "Preliminary" constitutes an exposition of the mathematical elements that will make possible the development of the science to come. From this Thesaurus of principles (more than a little suggestive, too, of the "topics" of Aristotle and the rhetorical tradition, as libraries of arguments to be brought to bear by the artist, as needed), Maxwell selects "Theorem III," which he proceeds to transform by mathematical artistry into the specific form of Thomson's theorem. This takes the form of an *identity*—not an equation valid only for certain values of the variables, called the "solutions," but an equivalence that holds true for any value of the variables whatever and thus permits the free substitution of one form for the other. Maxwell explains how he will use it:

> In Thomson's theorem, the total energy of the system is expressed in the form of the integral of a certain quantity extended over the whole space between the electrified bodies, and also in the form of an integral extended over the electrified surfaces only. The equality of these two expressions may be thus interpreted physically. We may conceive the physical relation between the electrified bodies, either as the result of the state of the intervening medium, or as the result of a direct action between the electrified bodies at a distance. If we adopt the latter conception, we may determine the law of the action, but we can go no further in speculating on its cause. If, on the other hand, we adopt the conception of action through a medium, we are led to enquire into the nature of that action in each part of the medium. (*Tr* i/62–63)[10]

The two hypotheses, Maxwell goes on to say, are *mathematically equivalent*. They represent the same quantity, but they express it in such different mathematical forms that they suggest altogether different ideas. Thus this "rhetorical" difference—a difference of "form" without a difference of

---

[10] For readers who may he interested in the mathematical forms of theorems discussed in this article, I will make certain of them available in the footnotes. "Thomson's theorem," as presented by Maxwell in the *Treatise* but with the notation somewhat modernized, takes this form:

$$V \iint E_n \, dS = \iiint E^2 d\tau.$$

Here, on the left-hand side, E_n, is the normal component of the electric intensity vector, E, which has the magnitude and direction of the force per unit charge on a very (*continued overleaf*)

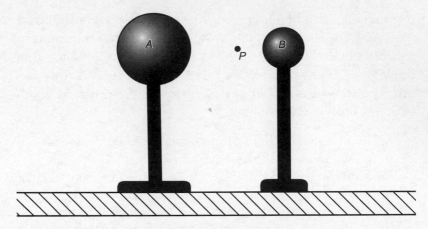

Figure 2 (repeated).

"content"—can make all the difference between two contrasting views of the world.

Thomson's theorem applies directly to our Figure 2. Here,
the two charged bodies repelling one another constitute a
system that is tending to fly apart. Like a wound-up spring, it
contains stored ("potential") energy. Thomson's theorem will
calculate that energy in two ways. The Old Way, taking the presumed charges on the surfaces of the spheres as fundamental,
will reckon these by means of a double-integration over the
surfaces of the two spheres (the double-integration signifying
a double-summation over the two dimensions of the surfaces).

The other side of Thomson's equation suggests the New
Way. Here the calculation of the same total energy of the system
is carried out over the region surrounding the charged bodies,
as a triple-integral representing a summation over the three
dimensions of that space. The resulting number, measuring

small test charge placed at a point in the region of a charged body.
V is the potential, which is uniform over the surface of a charged
conducting sphere; and dS is an element of that surface. On the
right-hand side, $d\tau$ denotes a volume-element of the space surrounding the sphere, and E is the magnitude of the intensity vector at any
point in the space.

the energy of the compressed system, is the same, reckoned by the Old Way or the New. But if we take the New Way, and sum over space, the question suggests itself immediately, "What is that quantity that we are summing?" It arose out of the mathematical identity, but we might find in it physical significance. Maxwell understands it as an energy density having a value everywhere in the region of the charged bodies, a "region" that has no boundaries. This is not an act of proof, nothing has been logically demonstrated; Maxwell's crucial suggestion, on which he founds field theory in the *Treatise,* is frankly a rhetorical act, one not of demonstration but of interpretation.[11]

The field arises, then, as the distribution of this energy density over space; it is the product of Maxwell's vision, and of his skillful mathematical rhetoric, which has shaped the equations so as to bring it to our view. We have entered, with Maxwell, a new conceptual world: what can we make of it?

The View from Point *"P"*

Maxwell distinguishes the Old Way and the New as the "direct" and the "inverse" methods:

> In the second chapter we have calculated the potential function and investigated some of its properties on the hypothesis that there is a direct action at a distance between electrified bodies, which is the resultant of the direct actions between the various electrified parts of the bodies.
>
> If we call this the *direct method* of investigation, the *inverse method* will consist in assuming that the potential is a function characterised by properties the same

---

[11] Here the act of interpretation consists in arbitrarily separating the integrand on the right-hand side of Thomson's equation, namely, E^2. When expressed in proper units, this quantity, now thought of as an energy density in space, becomes:

$$\omega = \tfrac{1}{8\pi} E^2 \text{ ergs/cm}^3.$$

as those which we have already established, and investigating the form of the function.

In the direct method the potential is calculated from the distribution of electricity by a process of integration, and is found to satisfy certain partial differential equations. In the inverse method the partial differential equations are supposed given, and we have to find the potential and the distribution of electricity....

The integral, therefore, is the appropriate mathematical expression for a theory of action between particles at a distance, whereas the differential equation is the appropriate expression for a theory of action exerted between contiguous parts of a medium. (*Tr* i/123–24) (italics added)

The term *differential equation* here envisions a statement, written in terms of derivatives, or rates-of-change of quantities (in electrostatics, rates-of-change in space rather than time), which characterizes the state of affairs everywhere in a region. Such an expression holds true in form at every point, though its specific value changes from point to point. For example, if we wanted to write such a differential equation to characterize the state of affairs in Figure 1, we would need to assert the fact that there are no charges in the space outside of spheres *A* and *B*. Maxwell shows how to write an equation (*Laplace's equation*), in terms of the electric intensity vector, that expresses this.[12]

This expression will be known to hold true throughout the field; the problem will be to find the actual values the intensity will have, everywhere, if specific values are set at the boundaries, namely, in this case, at the surfaces of the two spheres (the "boundary conditions").

---

[12] Laplace's equation:

$$\frac{\partial^2 V}{\partial x^2} + \frac{\partial^2 V}{\partial y^2} + \frac{\partial^2 V}{\partial z^2} = 0$$

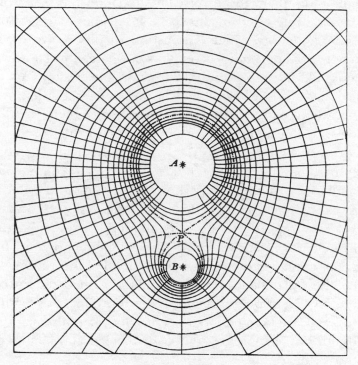

Figure 1 (repeated).

The resulting distribution of values at all points in the space is what Maxwell calls here "the form of the function" that is to be sought. Change anything—the value at a boundary, or the position of a charge at some point in the midst of the region—and the function will change form, all at once, everywhere. The inverse method sets a problem that can only be solved as a whole. In the direct method, applying Coulomb's law, we can mentally divide each charged surface into a large number of arbitrarily small elements and then total the effect of every element of the surface of sphere A, according to the law, upon every element of sphere B. Whereas, then, the direct method invites analysis, the inverse method demands a holistic approach; if such a problem is to be solved at all, it must be solved all at once.

Following, then, the inverse method into the new world, we become interested in the space itself, the field. Not only the energy density, which Thomson's theorem gave us, but its other properties must be explored; we need to know what it "looks like." It has a kind of directionality at all points, the directionality of the vector E. The inverse method delineates E as a whole pattern, a configuration. Tracing this configuration in connected curves, we get Faraday's lines of electric force, everywhere telling us the shape of the field. The corresponding lines of equal potential (analogous to contour lines on a hillside), running everywhere orthogonal to the lines of E, reveal mountains and valleys of complex shapes; Maxwell encourages us to utilize the "eye-knowledge" which diagrams of the field can give us, and he includes a number of these field diagrams as plates accompanying Volume I of the *Treatise*. The first of these plates illustrates the field that arises from the configuration of our spheres A and B; it is reproduced as Figure 1 of this essay. We would do well to consider what this image portends, for it is in a way a vision of the dialectical principle that motivates the *Treatise*.

Maxwell's figure shows us a *whole*, by contrast with the individual calculations that arise from action-at-a-distance formulas. If such a whole exists, then the question of the world-change becomes that of its significance. In the *direct* method, Maxwell said in the quotation above, "the potential is calculated from the distribution of electricity by a process of integration." "Integration" evidently means in this context a summing up—that is, the gathering of many elements into a totality. The whole is secondary, a consequence; it has the unity merely of marbles in a bag. Can we say that the inverse method carries an opposite significance?

Maxwell's figure, though it too is obtained by a process of integration ("integration" of the differential equation), is an image of continuity, bound everywhere by an equation that speaks of connections. "Integration" seems to have taken on a radically different meaning in the sea change between the direct and the indirect methods. In the first, the parts are

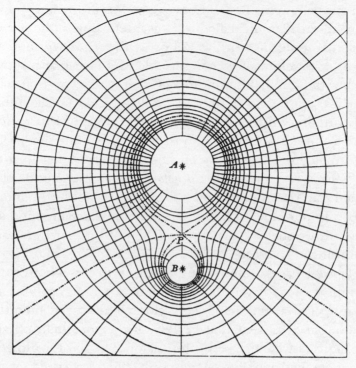

Figure 1 (repeated).

primary, and the integration sums them to find their combined effect. In the second, continuity is primary; the process
of "integration" in this case finds the single solution that will
satisfy the requirements of connectedness throughout. In the
first, the integral is a *summation of parts* to constitute a totality;
in the second, it is the *symbol of wholeness,* and the invocation of
a process in which parts will find their places with respect to a
whole that is primary.

Maxwell's close friend, Lewis Campbell, was a student
of Jowett's, and it is often noted that, as Maxwell went off to
Cambridge to immerse himself in mathematics, Campbell had
gone to Oxford to devote himself to Plato.[13]

---

[13] Campbell was Maxwell's original biographer: Lewis Campbell
and William Garnett, *The Life of James Clerk Maxwell.* The significance of his relationship with Maxwell is (*continued overleaf*)

Campbell opened a field of scholarship in his study of the dating of the Platonic dialogues; it is interesting in the present context to note his work as editor of the *Theaetetus* and the *Sophist*. Maxwell was very comfortable with Greek and kept in touch with Campbell's work to a certain extent over the years; there are reasons to think that he always had Greek distinctions running in the back of his mind. It might be in order, then, to invoke here the contrast which is so fundamental in those dialogues, the *Theaetetus* and the *Sophist*, between the *all* (*to pan*) and the *whole* (*to holon*). Socrates reminds us of Hesiod's account of a wagon: in one view, the wagon is made up of a hundred planks; in the other, it is that one thing, the *whole*, with respect to which alone any of the parts can have significance (*Theaetetus* 207a).

Maxwell was certainly not a stranger to this kind of consideration. He believed that the relation between physics and metaphysics was inherently very close. The names hint at it, he thought; it is clear that he felt much more than physical science itself was at stake in this major change of worldviews. The metaphysician, he says, is only the physicist disarmed of all his weapons: while the physicist measures time, space, and mass, the metaphysician struggles to speak of invariant sequence, coexistence, and the nature of matter. Physics has a certain priority; all serious thought must be grounded in consciousness, and hence the metaphysician would do well to enter upon his trade by way of physics.

It is hard not to glance again at the *Theaetetus* and consider the sense in which there is profound truth in Theaetetus's first response to the question, "What is knowledge?"—namely, "Knowledge is *aisthesis*"—which is of course the reason why it is appropriate that Theaetetus, perhaps the best of Socrates' students, comes to him from geometry, the science in which the senses and the mind join in argument. *Aisthesis,* commonly

---

emphasized by George Elder Davie: *The Democratic Intellect: Scotland and Her Universities in the Nineteenth Century.*

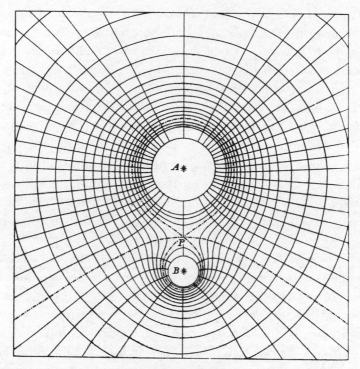

Figure 1 (repeated).

translated "sensation," is perhaps not so far from what Maxwell means by "consciousness"—though worlds separate the choices of the two terms. Maxwell praises, as we have seen, that "eye-knowledge" which his geometrical figure brings to our understanding of the world.

It appears to me that the "inverse method," represented in Figure 1, expresses the inversion not only of mathematical physics and electrostatics but more generally of the liberal arts and the world which they reflect. For example, in a late essay on freedom, Maxwell in effect takes the configuration of this diagram as an image of a new understanding of the operation of the free will.[14]

---

[14] "Does the Progress of Physical Science Tend to Give Any Advantage to the Opinion of Necessity … ?," Campbell and Garnett, op. cit., pp. 357ff.

He has wondered aloud whether the development of the physical sciences in the nineteenth century is leading necessarily to an increasing sense of determinism; he concludes that it need not, precisely because of points such as "*P*" in the figure. These are *singular* points, watersheds at which the least force could take us into one valley or the other. No force is required at such a point to change worlds.

The development of the principle of the conservation of energy, Maxwell says, has transformed our idea of the soul: it is no longer the source of motion, as it had been for Aristotle, but, rather, only the guide. He likens this role to that of the "pointsman" (the switch operator) on the railway, who turns the switch points to one side or the other and thereby deflects the course of a train to one or another destination. A simple configuration like that of our figure may have only one such saddle point, but as we go up the scale of being toward greater complexity, such points of indeterminacy multiply. Maxwell sees them as moral turning points, moments of tact, delicacy, and human insight which make crucial differences in our lives. "All great results produced by human endeavor depend on taking advantage of these singular states when they occur," he believes.

`Maxwell's notion of singularities could be a rather trivial suggestion in relation to the serious issue of freedom of the will, if it were no more than the observation that points of indeterminacy arise in physical systems. A little reflection on Newton's or Coulomb's law could lead quickly to the same conclusion. I think, however, that the distinction between the merely mechanical concept of instability and Maxwell's insight into free will helps us to see the full significance of the figure and the larger world-change that accompanies the shift from the old to the new view of the integral. For an instability, seen only as an ambivalent outcome of the application of a force law, gives us uncertainty without significance. On the other hand, the point *P* of the figure, seen as a saddle point between two valleys, has significance precisely because the individual decision point is not a point of blind indeterminacy but a point of vision, in which options can be comprehended

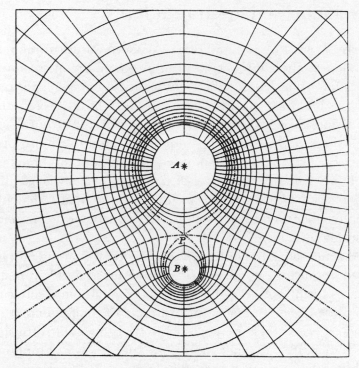

Figure 1 (repeated).

in their wholeness and relationship. That I think is an interesting insight into free will and is even exemplified in the *Treatise* itself. Rather than teach only his own method, Maxwell seems anxious to probe both the old and the new ways, and insofar as possible to understand them in their significance at this human turning point.

The *Treatise* is a dialectical work, which in its own way is pressing the questions of meaning and being. In insisting on a mode of physics that achieves intelligibility as fully as possible, Maxwell is pioneering in the effort to restore meaning in modern life and to help us understand what is at stake in the threat of its loss. In this sense, the *field* is for Maxwell a paradigm not only for the science of electromagnetism but for the method and structure of science generally, and for human thought, action, and society. The *Treatise*, I think, tries, as fully as a work of physics can, to achieve the view from point *P*.

Revelation and Interpretation

Before he went on to Cambridge, Maxwell was for three years a student at the University of Edinburgh, and many commentators have pointed out the significance that this experience had for his work. In Maxwell's time Edinburgh held firmly to a vision of liberal education entirely different from that at Cambridge. For an honors student at Cambridge the discipline was severe and single-minded. If a student was preparing for honors in mathematics—and this, especially under the influence of William Whewell, was the principal Cambridge track in liberal education—there was little time to spare for other matters.

At Edinburgh, by contrast, the program was broad and balanced, requiring attendance of all students at a wide range of lectures. In no case was the standard of attainment set at the Cambridge level; in place, for example, of total command of Greek, it was rather the goal at Edinburgh to learn through Greek to master language more generally. It was an Edinburgh tradition that lectures be accessible for general questioning in smaller groups following the lectures. In general, it was an education for men as human beings, not as specialists.[15]

The result was that for many students, of whom Maxwell was certainly one, the experience at Edinburgh left a very strong impression and a lifelong commitment to the pursuit of certain fundamental human questions. The concern with determinism and free will discussed above was certainly one of Maxwell's Edinburgh inheritances.

One center of Maxwell's attention at Edinburgh was Sir William Hamilton (not to be confused with the mathematician William Rowan Hamilton of Dublin, who in a very different way was also an important influence on Maxwell.) The Edinburgh Hamilton was famous for his courses of lectures, which all students attended, on logic and metaphysics. Many of

---

[15] Davie, op. cit., especially Part 3.

the recurrent themes in Maxwell's writings and in his scientific practice are recognizable as Hamiltonian concerns.

Hamilton was, in certain respects at least, a Kantian. At the foundation of his thought lies the principle of "Hamiltonian Relativity":

> ... the great axiom, that all human knowledge, consequently that all human philosophy, is only of the relative or phaenomenal ... in saying that we know only the relative, I virtually assert that we know nothing absolute.[16]

Our minds, on the other hand, are powerful forces that work upon the phenomena according to laws of the mind's own. There arises the threat of a tyranny of reason, against which Hamilton warns in many ways, for it is a fearsome temptation on the part of the mind, he thinks, to force consciousness into patterns of its own and to proclaim these to be truth. Hamilton particularly suspects philosophical "systems"; he speaks of the "Valhalla of Systems." What then is the right stance toward consciousness? He makes a bold claim:

> ... philosophy, as the science of truth, requires a renunciation of prejudices.... In this, if I may without irreverence compare things human with things divine, Christianity and Philosophy coincide—for truth is equally the end of both. What is the primary condition which our Saviour requires of his disciples? That they throw off their old prejudices, and come with hearts willing to receive knowledge and understandings open to conviction.... Philosophy requires an emancipation from the yoke of foreign authority, a renunciation of all blind adhesion to the opinions of our age and country, and a purification of the intellect from all assumptive beliefs.

------

[16] Sir William Hamilton, *Lectures on Metaphysics and Logic*, vol. i (Metaphysics), p. 96.

And then he draws an especially striking proposal:

> Consciousness is to the philosopher what the
> Bible is to the theologian. Both are revelations of the
> truth—and both afford the truth to those who are
> content to receive it, as it ought to be received, with
> reverence and submission.[17]

This passage points the way to my own thought. It is
Hamilton's vision of consciousness as revelatory that I propose as a clue to Maxwell's devotion to Faraday, as well as to
his own practice of science. I think Maxwell saw in Faraday the
"child," who was ready to purge himself of all prejudices and to
approach nature with the conviction that he was in the immediate presence of truth, which he was ready to receive with a
full measure of reverence and submission. Hamilton himself
goes on to a yet bolder assumption:

> ... I am ... bold enough to maintain, that conscious-
> ness affords not merely the only revelation, and only
> criterion of philosophy, but that this revelation is
> naturally clear.[18]

Consciousness is the equivalent of a sacred text—perhaps it
is a sacred text—and it will yield truth if we approach it in the
appropriate manner. Hamilton may be a "Kantian," but this is
not altogether Kant!

Faraday, as we have seen, does not conceive of science in the
mode of "theory," which appears to him rather an affront than
a model. These quotations from Hamilton, I believe, suggest a
point of view from which we might understand Faraday's position and, at the same time, Maxwell's. For if the task of science is
to be likened to that of theology in the presence of a sacred text,
the mode appropriate to science will not be theory but *interpretation*. Interpretation is a branch of the classic art of rhetoric; it

---

[17] Ibid., i, p. 58.
[18] Ibid., i, p.185.

was always important, but it became central with Augustine as the Greek art was brought into the service of Christianity. For Augustine, who most forcefully recast the ancient art in its new role, interpretation *(hermeneuein,* giving us "hermeneutics") was the art that made it possible to read and understand the voice of God in scripture, and indeed, as we see in the *Confessions*— which are an *interpretation* of a life—to "read" the Creation itself as well. All things for Augustine speak of their maker, for one who is able to hear and interpret their message.

Not only Augustine thinks this way. Maxwell's understanding of science—his dedication to interpretation as a way of bringing us closer to nature—is likewise suggestive in spirit, at least in my view, of Heidegger's invocation of hermeneutics in the approach to being, and thus I think it has not been inappropriate to suggest, as I did earlier in this essay, that we meet here a *hermeneutic,* as opposed to a *theoretic,* understanding of science. The field is thus an image of the hermeneutic view of science.

I should say, then, that Maxwell speaks with the Augustinian spirit in *interpreting* the meaning of the inverse method, which he sees as an inversion of our understanding of the cosmos:

> The vast interplanetary and interstellar regions will no longer be regarded as waste places in the universe, which the Creator has not seen fit to fill with the symbols of the manifold order of His kingdom. We shall find them to be already full of this wonderful medium; so full, that no human power can remove it from the smallest portion of space, or produce the slightest flaw in its infinite continuity. It extends unbroken from star to star.... (*SP* ii/322)

Alexandre Koyré has shown how the Copernican transformation led man out of the "closed world" into the "infinite universe."[19]

---

[19] Alexandre Koyré, *From the Closed World to the Infinite Universe.*

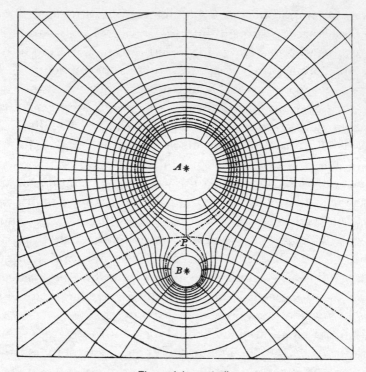

Figure 1 (repeated).

Newton found the key to interrelation of the elements of
the cosmos in the universal law of gravity, but the vast spaces
remained void, mediated only by an omnipresent God who is
evidently not the God in whom Maxwell places his faith. The
question does at this point touch on theology. Newton's God
is severe, the "God of power, lordship, dominion"—a God for
whom the autocracy of law, and the kind of dark intelligibility law brings, is appropriate. Maxwell has, in his mind, a very
different cosmos, a very different God, and a very different
concept of intelligibility.

The implications of the "inversion," in Maxwell's interpretation, are pervasive. The integrity, the wholeness we have spoken
of in connection with Figure 1, becomes the integrity of the
cosmos. The prospect is of a return from the "infinite universe"

to the restoration of the ancient cosmos as an ordered, contiguous whole. From the infinite universe, then, to the restored world. Figure 1, so interpreted, becomes the first image of that restoration.

Finally, since we form our human societies on the pattern of the cosmos and vice versa, the restoration of wholeness in the larger world suggests a similar restoration in the society of man. At this point, Maxwell's revolution in the realm of the cosmos has, as its implicit counterpart, that revolution from an "atomic" society to one of contiguity and membership of which Marx writes in the *German Ideology*. Indeed, it is appropriate that the careers of Maxwell and Marx seem to have had one fleeting point of tangency, in a common teacher of mechanics, Robert Willis. Willis's algebra of mechanics has its reflection in Maxwell's chapter "On the Equations of Motion of a Connected System," and in Marx's massive chapter in *Capital* on the analytic principle in capitalism, "Machinery and Modern Industry."

The two counterpart chapters reflect the two readings, physical and social, of our Figure 1. But to pursue this thought would require another investigation.

Analogy as an Instrument of Interpretation

We have seen, in Thomson's theorem, how far a mathematical figure of speech could take us in transforming a world. This transformation came in the early chapters; the *Treatise*, however, works constantly with another rhetorical instrument as well, that of *analogy*, which offers such versatility that it serves in one way or another at every level of investigation. At one extreme, it verges into analytic mathematics itself, as mathematical analogy; at the other, it delivers us to the doorstep of explicit physical theory. Maxwell uses it, not only in this work but in the papers that led up to the *Treatise* and in the books that arose from it, with a special appreciation of the intellectual and strategic alternatives it affords him. He sometimes stops to discuss the role of "analogy" with the reader. Employment of what we might call the right grade of analogy makes it possible to meet the rhetorical criterion Maxwell holds as a standard: to assert all that is justified by the evidence, but no more.

Maxwell returns, in chapter 5 of Part I of the *Treatise,* to ask, from a more physical point of view, what sort of medium may in fact exist in the space between *A* and *B*? This will not be a specific physical hypothesis, nor a physical *theory* of the field. All that Maxwell explores is the possibility that an analogy can be constructed, by virtue of which the phenomena of electrostatics can be systematically *likened* to those of a physical medium under stress. In this way, our thought is in a way empowered, but in a way it is also suspended. We realize this if we ask just what sort of "medium" we seem to be getting at and look closely at Maxwell's chapter to see what details he gives us. He gives us none; there is a strong suggestion he is playing a kind of game with us, knowing we will be wondering, and saying quite strictly nothing to help us out of our distress. Is this a solid undergoing elastic deformation? He will not say so but only repeats that word, *medium.* It is *like* an elastic solid, and he can use this strict, quantitative analogy to carry out a calculation of the stresses.

The calculated stresses within this metaphoric medium prove to be these: first, a tension along the line of force, which can thus be thought of as stretched like a rope; and second, a stress orthogonal to the line and equal in magnitude to the first. It is evidently a matter of great satisfaction to Maxwell to be able to show us that this is exactly the vision which Faraday had given us. Maxwell cites Faraday's account, and concludes on a note of quiet triumph:

> This is an exact account of the conclusions to which we have been conducted by our mathematical investigation. At every point of the medium there is a state of stress such that there is tension along the lines of

force and pressure in all directions at right angles to
these lines, the numerical magnitude of the pressure
being equal to that of the tension, and both varying as the square of the resultant force at the point.
(*Tr* i/164)

Having won this much ground, Maxwell now undertakes a
Circe-like trick which, as a work of rhetoric, may be more disturbing than the inversion of worlds. He returns to the concept
with which the whole science of electrostatics began, the concept on which the science is founded, and makes it disappear
before our eyes. He shows that "charge" was a work of what
Lukács would call "reification," an object of false consciousness
bred of living too long in a wrong world.[20]

*There is no need for the concept of "charge" in the new world of
Maxwell's fields.* There is no "center of force" anywhere. In its
place, we will have the very different concept of electric *displacement* throughout the field. Using our new image of an
elastic medium, we may consider what happens when a positively
charged sphere is placed in the center of a room. The air proves
to be *dielectric,* for through the air corresponding apparent negative charges are induced on the walls, floor, and ceiling. With
care, we could trace the lines of force from the sphere to those
surfaces. With our new insights we could tell ourselves that there
is an electric stress everywhere, which Maxwell likes to evaluate
in pounds per square inch. This is an instance of induction: the
charged sphere, Faraday has said, is *inducing* the wall charges
through the dielectric medium of the air.

In this context, Maxwell now introduces his term *displacement*:

When induction is transmitted through a dielectric,
there is in the first place a displacement of electricity

---

[20] Georg Lukács, "Reification and the Consciousness of the
Proletariat," in *History and Class Consciousness*, pp. 83ff.

in the direction of the induction. For instance, in a
Leyden jar, of which the inner coating is charged
positively and the outer coating negatively, the direction of the displacement of positive electricity in
the substance of the glass is from within outwards.
(*Tr* i/166)

Applying this to our case of the charged sphere in the
center of a room, we would say: there has been a displacement of positive electricity from the sphere toward the walls.
Where, then, is the "charge," that supposed substance on
which the science of electricity is founded, and which is the
center of all forces? There is, after all, nothing deposited on
what have been called "charged" surfaces, but rather a state
of affairs, a state of displacement or polarization, throughout
the space:

> Thus when the charged conductor is introduced
> into the closed space [the positively charged sphere
> into the room, "bringing" electricity with it] there is
> immediately a displacement of a quantity of electricity
> equal to the charge through the surface from within
> outwards, and the whole quantity within the surface
> remains the same. (*Tr* i/68)

We *thought* we were "bringing" something called "electricity," or "charge," into the room on the "charged" conductor.
Actually, by the analogy, we are rearranging the state of affairs in
the dielectric medium which fills the room—placing it in a state
of stress through a displacement outward from the sphere, to,
and *through,* the walls. There is nothing "on" the sphere or "on"
the walls. The displacement, and the stress, would be essentially
the same if the room were emptied of its air and the experiment
were carried out in total vacuum. We have accounted in very different terms for "that apparent charge which is commonly called
the Charge of the Conductor." All of this is the consequence of
a method not of proof but of interpretation—of entertaining a
vivid physical analogy very seriously.

The Transformation Extended to Magnetism

We have followed Maxwell in the transformation of our understanding of electrostatics, but electrostatics is only half of the overall science of electricity and magnetism that Maxwell is addressing in the *Treatise*—and electrostatics is perhaps the easier half. Much more confusing phenomena arise in the domain of magnetism, and especially in relation to "electrodynamics," in which electric currents act magnetically upon one another. In electrostatics we had phenomena which seemed to fit the pattern of the gravitational force between the sun and a planet—Coulomb's law was an easy counterpart to Newton's law of universal gravitation. Maxwell's conversion of thought from Coulomb's law to the field point of view was indeed a delicate maneuver, but it represented a straightforward problem and serves well as a paradigm in which we can study the inverse method and consider its consequences. In electrodynamics, and more generally in the phenomena of magnetism, things are not so clear. Thus, the first electrodynamic phenomenon observed, by Ørsted, was the action of a straight wire carrying current upon a magnetic needle nearby: the poles of the needle were *neither attracted nor repelled* by the wire but were moved in a direction perpendicular of the line joining them to the wire. The effect was rotational, and Faraday in fact showed how to use it to produce a continuous rotational motion. Surely this could not be fitted to the Newtonian model!

It was a moment of special satisfaction for mathematical physics when Ampère succeeded in doing just that—in showing that an inverse-square law could be written which, when applied pair-wise to arbitrarily small current elements, or snippets of current, would account for the observed torque. If mathematical physics were coextensive with laws of force acting at distances along a line of centers, as it seemed, then Ampère had saved the day for mathematical physics, and he was widely celebrated for this victory.

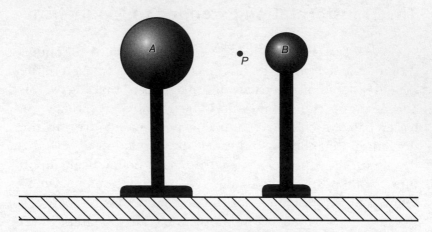

Figure 2 (repeated).

Again, to "fix our ideas" we may think of a particular electrodynamic situation that is in a way a counterpart to the electrostatic arrangement of Figure 2. Here (Figure 3), two wires are carrying current; for simplicity, we may think of them as lying in two parallel planes. Between them will occur a magnetic interaction—an attractive force of one wire upon the other if the currents flow as shown in the diagram. Ampère's law applies to two arbitrarily small segments of the two circuits, so-called current elements, ds and ds'. The law predicts a force along the line r joining them, given by a rather complex formula—not at all like the straightforward equation of Coulomb.[21]

The very arbitrariness and awkwardness of the formula makes it suspect in Maxwell's eyes: it seems rather imposed on nature than drawn from it. Furthermore, the putative "force" between ds and ds' can never be demonstrated empirically, for we cannot produce a current in a segment like ds: currents by

---

[21] Ampère's formula involves the angles θ and θ', which are the angles that the current-elements make with the line which joins them, whose length is r, and the angle ε, which the two current-elements, translated so that they met at a common point, would make with each other. In these terms the equation is:

$$df = \frac{i\,i'}{r^2}\left(\cos\varepsilon - \tfrac{3}{2}\cos\theta\,\cos\theta'\right)ds\,ds'.$$

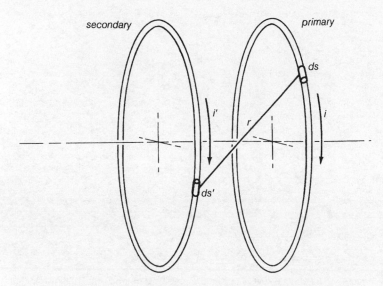

Figure 3: Two circular coils carrying current in parallel planes. If the currents, *i* and *i'*, are flowing as shown, the force between the coils will be attractive; *ds* and *ds'* denote the current elements, and *r* the distance between them, in Ampère's law. If a switch is introduced in the primary, and a galvanometer in the secondary, electromagnetic induction can be demonstrated.

their nature must flow in complete circuits. Ampère's equation must therefore be summed around at least one of the circuits before it can be used, and with that summation, the supposed force along the individual line *r* is completely swallowed up. Thus, the "force" of which Ampère's law speaks can never be observed—and that, in Maxwell's eyes at least, is a very bad sign for any proposed law.

Faraday, of course, takes an utterly different approach to this interaction. We will look at his discussion of the interaction of magnets, rather than the currents of Figure 3, but as the interactions are magnetic in either case, what he says will be applicable to electrodynamics as well. Again, Faraday thinks in terms of events transpiring in the space between the magnets, and again, he delineates this region, the field, in terms of lines of force. If we think for a moment of an isolated magnet, the whole system of its lines of force constitutes for Faraday what he comes to see as an entity, the real locus of the magnetic

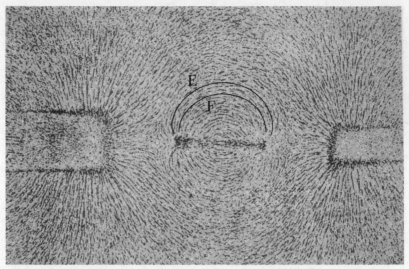

Figure 4. The sphondyloid of power of a small bar magnet, caught between the sphondyloids of two larger magnets. Iron-filings diagram adapted from Plate V of Faraday's *Diary*.

power, a kind of "atmosphere of power" with a definite, though alterable, form (Figure 4). To designate this spatial object, he again calls on "the advice of a kind friend" to coin an appropriate name. Inasmuch as here the lines suggest to him the form of a winged beetle, the word made to his prescription is *sphondyloid,* or beetlelike (Greek: *sphondylon,* beetle). He defines the newfound *sphondyloid of power* in this way:

> If, in the case of a straight bar-magnet, any one of these lines, *E,* be considered as revolving round the axis of the magnet, it describes a surface; and as the line itself is a closed curve, the surface will form a tube [torus] round the axis and inclose a solid form. Another line of force, *F,* will produce a similar result. The sphondyloid body may be either that contained by the surface of revolution of *E,* or that between the two surfaces of *E* and *F*.... (*XR* ¶3271)

Faraday has ways of tracing these lines, and even of measuring the power of the total sphondyloid of a given magnet.

He comes to view magnetic actions as taking place primarily between sphondyloids, and only secondarily between the

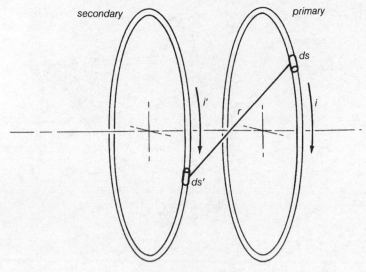

Figure 3 (repeated).

magnets: Thus, as we earlier dispensed with "charge," so now we eliminate the magnetic "pole." Interactions between magnets take on the proportions of Homeric encounters:

> How easily all these effects present themselves in a consistent form, if read by the principle of representative lines of force! ... as the [spherical] magnet is approached [by the dominant], its external sphondyloid of power is compressed inwards ... and at last the magnet is self-contained ... so that it gives no induced currents.... Within that distance the effect of the superior and overpowering force of the great magnet appears ... which, though it can take partial possession of the little magnet, still, when removed, suffers the force of the latter to develop itself again .... (*XR* ¶3359)

Maxwell, in transforming the mathematics of magnetic interactions from Ampère's form to a field point of view, will take Faraday's depiction of interacting sphondyloids as a guide to the kind of account he will need to produce.

As in the electrostatic case, Maxwell summons a mathematical figure of speech to his aid; we cannot here follow him in this but only report the outcome: the currents of Figure 3 can

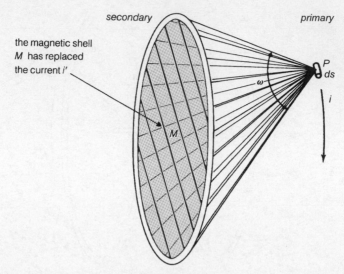

Figure 5. The Gauss potential of a circuit. The secondary current of Figure 3
has been replaced by a magnetic shell, *M* (north face on the right, south on
the left); and from the position of *ds* of Figure 3, now denoted *P*, the shell
delineates the solid angle *ω*, which measures its potential.

be shown to be mathematically equivalent to magnetic shells
of which they are the boundaries. A *magnetic shell* is an imagi-
nary surface which is magnetized overall in such a way that one
face corresponds everywhere to the north pole of a magnet,
and the opposite face to the south. In turn, Maxwell borrows
from Gauss a powerfully intuitive proposition concerning the
potential of a magnetic shell, *M*: the potential at any point *P* is
proportional to the solid angle subsumed by the boundary of
the shell when seen from that point (Figure 5). The effect of
this combination of strategies is to escape completely from the
analytic frame of mind of Ampère, and to conceive the magnetic effect as that of the circuit as a *whole*. The mathematical
concept of potential catches the spirit of Faraday's imaginative grasp of the magnetic phenomenon as a power in space.
By extending the approach of Figure 5 to include the entire
region of the magnet, Maxwell is able to grasp the field in the
case of magnetism very much as he did in the case of electrostatics, and to conceive again of the whole configured field as
incorporating a single potential of the circuit.

Figure 6. Adaptation of Maxwell's diagram of the magnetic fields of two circular currents (Figure XIX of Volume II of the *Treatise*). This figure represents a vertical section through the circuits of Figure 3. Figure 6 superimposes on Figure 3. The symbol ⊙ represents currents flowing out of the page; symbol ⊗ represents currents flowing into the page. Lines marked with arrows, together with those running alongside them, represent lines of force. Equipotential lines run everywhere perpendicular to these.

The lines of *force* run here, as they did in the electrostatic case, at right angles to the lines connecting points of equal *potential,* or equipotential lines: that is, if we think of the equipotential lines as gravitational "contour" lines, the lines of force run orthogonally to them, or "downhill." All this is depicted in Maxwell's drawing of the combined field of the two circuits of Figure 6. Again, Maxwell has captured in mathematical form the essence of Faraday's insight and has replaced the analytic,

formal thinking of Ampère with an integrated, synthetic, and
highly intuitive alternative. This is the same conversion of the
world we met in the case of electrostatics, now successfully
carried out in the much more complex and elusive realm of
magnetic interactions.

At the beginning of chapter 3 of Part IV of the *Treatise*,
Maxwell devotes three remarkable pages to a dialectical reflection on these two contrasting understandings of the nature of
"science." He draws here again the distinction we have seen
earlier between the two contrasting views of the *integral*: as a
sum of parts, or as a whole. Ampère, Maxwell says, builds a
complex by summation; Faraday begins with the whole and
derives the particular from it:

> We are accustomed to consider the universe as
> made up of parts, and mathematicians usually begin
> by considering a single particle, and then conceiving
> its relation to another particle, and so on. This has
> generally been supposed the most natural method.
> To conceive of a particle, however, requires a process
> of abstraction, since all our perceptions are related
> to extended bodies, so that the idea of the *all* that
> is in our consciousness at a given instant is perhaps
> as primitive an idea as that of any individual thing.
> (*Tr* ii/176–77)

Ampère shut out this wholeness of the act of perception by
a deliberate effort to construct a system on the concept of least
parts; thus his force law expresses an action between current
elements whose relation can never be observed as such in fact.
Faraday, proceeding in a way that seems to Maxwell the natural one, accepts what he observes in its immediate unity. He
sees magnets as wholes, he sees the patterns of lines of force
in their unity as whole systems, and he names that unity, the
"sphondyloid." By the necessities of our mode of perception,
a nascent science will be a science of such wholes. Ampère was
able to produce a science of another sort only by erasing the

record of his actual investigation and by forcing his science to conform to a rigid tradition. Faraday, by his freedom from this discipline, has been able to set his own style and shape his own language. The result is an ability, Maxwell says, "to coordinate his ideas with his facts," and to achieve their expression "in natural, untechnical language." Style is crucial to truth, because only a "natural untechnical" style will be responsive to nature in the manner of the phenomena themselves. Maxwell says of Faraday:

> This new symbolism consisted of those lines of force extending themselves in every direction from electrified and magnetic bodies, which Faraday in his mind's eye saw as distinctly as the solid bodies from which they emanated. (*SP* ii/318)

In this context, we see how much the metaphor of "the mind's eye" carries with it. As the eye immediately grasps the *whole* in perception, so the mind's eye intuits the concept of field as an extended entity, the system of the lines of force. The field is thought of here essentially as a *whole,* as an entity. Field physics must be close to our immediate intuition of nature, if Maxwell is right about the character of perception.

Electromagnetic Induction: from Statics to Dynamics

Let us turn once again to the phenomenon of electromagnetic induction. When a current flowing in one circuit, the "primary," is started or stopped, a current will be induced in another neighboring "secondary" circuit. In the field view, it is the changing magnetic field of the first circuit that induces the current in the second. But, if a changing field induces currents in other circuits, it should do so in the original circuit as well. This is the phenomenon of "self-induction." Furthermore, by the conservation of energy, the induction must be in a direction to oppose the original change; otherwise, we would have an explosive regeneration of energy. This, too, is in accord with observed phenomena: when a switch in the primary circuit is opened in order to stop current flow, an arc may occur across the points of the switch, the current resisting with a high potential the interruption of its flow.

Maxwell observes, in chapter 4 of Part IV, that this has an inescapable analogy to the flywheel effect in mechanics: the momentum by which a moving mass resists change in its motion. The suggestion becomes very strong that there is, associated with the current, some counterpart to mechanical momentum, which Maxwell comes to call "electrokinetic momentum." It is not literally a momentum of "electricity," for it varies with different geometrical arrangements of the wire (as it would not, for example, if we were speaking of water in a hose). The momentum must in some way be located, not in the wire but in the field. Maxwell's mind moves eagerly in response to this leading thought:

> It is difficult … for the mind which has once rec-
> ognised the analogy between the phenomena of
> self-induction and those of the motion of material
> bodies, to abandon altogether the help of this anal-
> ogy, or to admit that it is entirely superficial and
> misleading. The fundamental dynamical idea of mat-
> ter, as capable by its motion of becoming the recipient
> of momentum and of energy, is so interwoven with
> our forms of thought that, whenever we catch a
> glimpse of it in any part of nature, we feel that a path
> is before us leading, sooner or later, to the complete
> understanding of the subject....
>
> It appears, therefore, that a system containing an
> electric current is a seat of energy of some kind; and
> since we can form no conception of an electric cur-
> rent except as a kinetic phenomenon, its energy must
> be kinetic energy, that is to say, the energy which a
> moving body has in virtue of its motion....

We are therefore led to enquire whether there may
not be some motion going on in the space outside the
wire, which is not occupied by the electric current, but
in which the electromagnetic effects of the current
are manifested. (*Tr* ii/196–98)

Maxwell speaks cautiously, but he is already committed to a
perhaps extravagant adventure:

... Has the electric current ... either momentum or
kinetic energy?
We have already shewn that it has something
very like momentum, that it resists being suddenly
stopped, and that it can exert, for a short time, a great
electromotive force.
But a conducting circuit in which a current has
been set up has the power of doing work in virtue
of this current, and this power cannot be said to be
something very like energy, for it is really and truly
energy. (*Tr* ii/197)

The energy is "real and true," the momentum is "very like"—
fine distinctions of the art of interpretation! We shall see where
they lead Maxwell's searching mind.

A New Mechanics for a New World

This will be a metaphorical journey. When Maxwell says that
we have observed "something *like* momentum," we sense that
an analogy is brewing. Maxwell has used analogies extensively
before: mathematical analogy in his first paper on electricity,
"On Faraday's Lines of Force"; physical analogy in the second
paper, "On Physical Lines of Force"; and both kinds in various ways earlier in the *Treatise* itself. Now he is going to travel
further with analogy than he has ever done before. Previously,
metaphor bridged two systems that were somehow alike. The
metaphor spoke of the abstracted likeness, but at each end

the cable suspending thought was anchored in *terra firma*. Now
Maxwell will propose instead that the suspended thought itself is
the reality: from *terra firma* at either end we raise our thoughts to
terra nova in the intellectual space between. Maxwell has found
energy, and something like momentum, in the magnetic field in
a pure vacuum: where "nothing" was, there is a thinkable mechanism, running and charged with momentum and kinetic energy.

Here is a kind of being that is real and material, but invisible, intangible, and imponderable. Can we open our thinking
to permit the possibility that "matter" is not what we had
thought, that very real "matter" could be present in this empty
space? Maxwell is indeed asserting the total reality of the field,
yet he is not reducing it to a "mechanical system" in the way
in which nineteenth-century physicists are accused of having
naively tried to do. Rather, he is using metaphorical leverage
to raise our thinking to a new level of understanding of what
physical reality might entail. I am not sure his critics have fully
understood him in this.

Here is Maxwell's announcement of his project:

> What I propose now to do is to examine the con-
> sequences of the assumption that the phenomena of
> the electric current are those of a moving system, the
> motion being communicated from one part of the
> system to another by forces, the nature and laws of
> which we do not yet even attempt to define, because
> we can eliminate these forces from the equations of
> motion by the method given by Lagrange for any con-
> nected system.
>   In the next five chapters [chapters 5–9, Part IV, of
> the *Treatise*] I propose to deduce the main structure of
> the theory of electricity from a dynamical hypothesis
> of this kind…. (*Tr* ii/198)

To carry out this project will require a highly sophisticated
instrument, and Maxwell found what he wanted in what are
known as the "Lagrangian equations of motion." This is a set

of partial differential equations, which, though published by Lagrange in 1788 in his *Mécanique analytique,* were only becoming widely known in England during the time in which Maxwell was working on electricity and magnetism.[22]

They describe in purely analytic terms the motions of a connected mechanical system. The emphasis is on connection, because one special interest of the equations lies in the fact that they eliminate inner connections from explicit consideration. They thereby leave a set of truly independent, "generalized" coordinates with which to describe in observable external terms the state of an inwardly, fully connected system. Any sufficient set of strictly independent coordinates will do. They may be Newtonian displacements, but they need not be—they could be the angular displacement of a knob, the positioning of an odd-shaped cam, or the pull of a string. These are Lagrange's "generalized coordinates," and their time derivatives will be correspondingly "generalized" velocities. We are embarking on a sea of metaphoric displacements, velocities, and their offspring: in Lagrange's equations, they will *act like* their Newtonian counterparts, but they need not *be* the kind of thing Newton was referring to in his laws.

Seen from Maxwell's point of view, Lagrange's equations are sparkling with interest and possibilities, for they begin with the concepts of *connection* and *system*; they look to the *whole,* which they characterize in terms of the scalar energy, and by identifying a set of variables of any sort that is just sufficient to determine a *state* of the system, they make it possible to meet the interpretative ideal of saying exactly what is known and no more.

Maxwell described Lagrange's equations in a parable which is very helpful in understanding their significance to him from what we might call an epistemological point of view. He

----

[22] They were popularized especially by Maxwell's friends William Thomson and P. G. Tait, in their *Treatise on Natural Philosophy.*

envisions a group of bell ringers pulling upon ropes connected
to a set of bells inaccessible overhead:

> In an ordinary belfry, each bell has one rope which
> comes down through a hole in the floor to the bell-
> ringers' room. But suppose that each rope, instead of
> acting on one bell, contributes to the motion of many
> pieces of machinery, and that the motion of each
> piece is determined not by the motion of one rope
> alone, but by that of several, and suppose, further,
> that all this machinery is silent and utterly unknown
> to the men at the ropes, who can see only as far as the
> holes in the floor above them.
>
> Supposing all this, what is the scientific duty of the
> men below? They have full command of the ropes, but
> of nothing else. They can give each rope any position
> and any velocity, and they can estimate its momentum
> by stopping all the ropes at once, and feeling what
> sort of tug each rope gives. If they take the trouble to
> ascertain how much work they have to do in order to
> drag the ropes down to a given set of positions, and
> to express this in terms of these positions, they have
> found the potential energy of the system in terms of
> the known co-ordinates. If they then find the tug on
> any one rope arising from a velocity equal to unity
> communicated to itself or to any other rope, they can
> express the kinetic energy in terms of the co-ordinates
> and velocities.
>
> These data are sufficient to determine the motion
> of every one of the ropes when it and all the others
> are acted on by any given forces. This is all that the
> men at the ropes can ever know. If the machinery
> above has more degrees of freedom than there are
> ropes, the co-ordinates which express these degrees
> of freedom must be ignored. There is no help for it.
> (*SP* ii/783–84)

Maxwell, who very much enjoyed a quiet joke, must have
taken some satisfaction at the thought of the music that might
result.

In what we might call its first level of application, the Lagrangian method is an illustration of the ultimate interpretive instrument in that aspect of the art in which precision of expression is the goal, for it is capable of expressing "all that the men at the ropes can ever know," and no more. By employing just those variables that are accessible and can be measured and controlled, we are able to speak of just what is strictly accessible to us. In this, Lagrange's equations come close to the controlling rhetorical principle of the *Treatise*:

> We must therefore seek for a mode of expression which shall not be capable of expressing too much, and which shall leave room for the introduction of new ideas as these are developed from new facts. (*Tr* ii/7)

At a deeper level of application, the method not only permits a disciplined expression in familiar terms but opens the way to an advance in our understanding through the reformation of the fundamental terms themselves. The electromagnetic field will of course be the machinery of the bells that concerns us. The bell ropes and their velocities become galvanometer deflections and their motions, or the position or motion of a wire near a magnet. We shall be led to introduce a whole new metaphoric language, in which we speak of "electromotive force," "electric current," "electric displacement," or "electrokinetic momentum." In a sense, we have no notion initially what any of these things is. But we may enlarge our notion of force so that it is not a misnomer, or a "mere analogy," to speak, for example, of the electromotive force. Possibly a "force" is something more than we thought.

Perhaps the most difficult concept to deal with on this metaphorical level is "matter" itself, and I would like to take a moment to review some of Maxwell's thoughts about this, and the thoughts, too, of his teacher, William Hamilton of Edinburgh, who I think may help us. Let me start with Hamilton, whom we have already met earlier in this essay. Hamilton insists on the duality of all perception; we invariably,

he says, find mixed in our consciousness facts of two kinds—
"internal facts" and "external facts." The former testify to an
inner reality, the ego, and the latter to an external reality, the
non-ego. This duality of perception is ignored, Hamilton com-
plains, by philosophers who are determined to achieve unity
and simplicity by the elimination from consideration of one
or the other of these components. Embracing both in his
"Philosophy of the Conditioned," Hamilton commits himself
to what he calls "natural dualism," or "natural realism."

"What is meant," Hamilton asks, "by perceiving the material
reality?" He is prepared to answer in terms of "dualism":

> In the first place, it does not mean that we perceive
> the material reality absolutely and in itself … on the
> contrary, the total and real object of perception is the
> external object under relation to our sense and faculty
> of cognition. But though thus relative to us, the object
> is still … the non-ego—the non-ego modified, and
> relative, it may be, but still the non-ego…. Suppose
> that the total object of consciousness in perception
> is = 12; and suppose that the external reality con-
> tributes 6, the material sense 3, and the mind 3—this
> may enable you to form some rude conjecture of the
> nature of the object of perception.[23]

Hamilton is willing to assign certain primary qualities to
external reality, one of these—*pace* Kant—being "extension."
How can "extension" be attributed to external reality, while
"space" is asserted (with Kant) to be strictly a condition of
our consciousness? Hamilton pauses for a long breath before
answering that one:

> To this difficulty, I see only one possible answer. It
> is this: It cannot be denied that space, as a neces-
> sary notion, is native to the mind; but does it follow,
> that, because there is an *a priori* space, as a form of

---

[23] Hamilton, op. cit., i, p. 357.

thought, we may not also have an empirical knowledge of extension, as an element of existence? The
former, indeed, may be only the condition through
which the latter is possible…. But there seems to me
no reason to deny, that because we have the one, we
may not also have the other. If this be admitted, the
whole difficulty is solved….[24]

Hamilton seems to have opened the trapdoor in the ceiling
halfway, to glimpse the forbidden things-in-themselves, and to
find that there is thus some communication between thought
and fact.

In the image of the bell ringers, we can identify the three
components of Hamilton's recipe for matter. The six portions
of external reality lie sealed from me above the ceiling. Three
portions are the laws of operation of my own reason, exercising
my "scientific responsibility" to make what I can of the situation. But an interesting three portions of material sense are
conveyed by the bell ropes, which communicate between the
two realms. Lagrangian theory is a strategy of reason for interpreting the information of the material sense: I think we can
say that, in expressing the idea of a connected material system, it mirrors the conditions of reason itself. The principle
of relativity tells us that we can know only the configuration of
the ropes among themselves, each with respect to the others.
There is no absolute reference by which we could measure
them more truly. The Philosophy of the Conditioned tells us
that we can never assert what is true, simply. But the recipe
for matter gives us reason not to despair: we do have a dash of
contact with being, and that, I believe, for Maxwell makes it all
worthwhile.

There are, Maxwell told his students in experimental
physics, two "gateways to knowledge": one by way of the "doc-
trines of science," the other by way of "those elementary
sensations which form the obscure background of all our

---

[24] Ibid., p. 346.

conscious thoughts." If it is possible to "effect a junction in the citadel of the mind, the position they occupy becomes impregnable." To make this junction is to "wed Thought to Fact"; to effect a mystery. This might be the key to both the content and the style of the *Treatise*: it has dedicated itself to achieving that union insofar as is possible, by way of the emerging science of electromagnetism and the insights that new science may bring with it. The *Treatise*, I suspect, has more than what is ordinarily regarded as respectable scientific work to do.

Between "Thought" and "Fact," Maxwell sees one link as that of analogy. Analogy builds a bridge, then, between "matter" as sensed and "matter" as known to the mind. All this Maxwell discusses in the speculative essay, "Are There Real Analogies in Nature?"[25] Thought, he says there, deals in formal relations; whenever the opportunity arises, by a necessity of our natures, we project these relations upon phenomena, thereby *imposing* analogies. Is this only imposition, no more than a "mere projection of our mental machinery on the surface of external things"? I leave it to the reader to turn to Maxwell's essay to see what the answer might seem to be. It was an early essay; the advice to the students of experimental physics was much later. If that latter seems more confident, it may be because the experience in the interval with Lagrangian theory and the *Treatise* has raised Maxwell's spirits in relation to being.

It is significant that Maxwell has his own way of deriving Lagrange's equations in the *Treatise*. Where others, including Lagrange himself, anchor the derivation of the new forms in Newton's laws and Newton's concepts, Maxwell uses the terms of the new science from the beginning. We wake up, so to speak, on the magic island and hear the sounds of Ariel's music. Maxwell above all will not use Newton's fundamental term, *mass*. This is a kind of tacit joke on Newton, to steal his science by refounding it without the use of its primary concept.

---

[25] Reprinted in Lewis Campbell and William Garnett, *The Life of James Clerk Maxwell* (1882), p. 237.

Other terms, such as *coordinate, velocity, momentum, or force,* though familiar in sound, now are used with new meanings—a fact which may only gradually become apparent to the reader. We are indeed under some kind of spell, under a metaphoric charm. Nothing is quite what it seems, nor ever will be, for this is a dialectical turning point, and once a new idea is entertained the world cannot again be the same.

We cannot here trace the unfolding of Lagrange's equations under the guidance of Maxwell's magic, but we may note one or two high points. The beginning is with a statement that sounds like Newton's second law of motion, and Maxwell is no doubt confident we will accept it as such, but in fact it is referring to generalized coordinates and hence bears an altogether new meaning: "The moving force is measured by the rate of increase of the momentum" (*Tr* ii/201). Except for the relation between them, *force* and *momentum* remain undefined terms—we are, in effect, invited to form ideas of the generalized terms directly. In general, they are not Newtonian force and momentum.

Of the two quantities, "force" and "momentum," it is the former which is treated as intuitively prior, with the result that the edifice of mechanics is built on a new fundamental understanding of force itself. In turn, a generalization of the idea of force means a new approach to the concept of causality, and Maxwell seems indeed to invite this as well. Separating the sensations of effort or resistance from the dynamic idea of force, Maxwell urges us to consider that in dynamics the planets move "like the blessed gods."

Maxwell next introduces the concepts of work and energy, assuming initially that the work done by a force is the product of the force by the space through which it acts. But if force is generalized, then the space which multiplies it must be, as well. The concept of space is opened to generalization in company with the concept of causality. And at last, at the top rather than the base of the edifice, we must encounter mass, as transformed as "space" and "force" have been. It appears

merely as the coefficient by which the one-half of the square of
the generalized velocity must be multiplied to give the kinetic
energy. It bears a new name: not "mass," but the "moment" or
"product" of "inertia."

Strictly speaking, perhaps, nothing in the new system has
been defined. This might seem to represent resignation,
total acceptance of Hamiltonian relativity in its darker aspect.
I would rather remember the words about the way in which
"Thought weds Fact" and believe that the new mechanics is not
a pure formalism but an artfully constructed system of symbols
inviting meaning. And it would seem that the encounter with
electromagnetism to which we turn now should be the occasion of the epithalamion.

Maxwell, on behalf of the Society for Promoting Christian
Knowledge—surely not without the memory of Faraday very
much in his mind—prepared a primer of the new mechanics
as nearly as possible without recourse to formal mathematics.
Always, for Maxwell, the first test of intelligibility of a mathematical structure is restatement in honest prose. *Matter and
Motion*,[26] the little book to which I am referring, introduces
the new mechanics on an elementary level; "elementary,"
not in the sense of "easy," for with a little practice analytic
mathematics is always the easiest way to go, but in the old and
important sense of "elementary." In this book Maxwell states,
as accurately and adequately as prose and his skill permit, the
real foundations of the new understanding, and in the course
of this book he works through to the question which to him
is most burning: the argument to Newton's law of universal
gravitation. For Maxwell now knows that this law cannot be
true. The book is not a diluted account for the nonscientist;
rather, it brings the adult reader, scientist or not, to stand with
Maxwell and join him in his thoughts as he stares into the heart
of the real question, the unresolved prospect that lies ahead:
The Twentieth Century, I think it might be called.

---

[26] Society for Promoting Christian Knowledge (1876). Reprinted
Dover Publications (1991).

The Procession of Maxwell's Equations from Lagrange's

Maxwell now breathes life into Lagrange's equations by bringing them to bear on Fact—on the phenomena of electromagnetism. Lagrange's equations presuppose that we are dealing with a single, connected material system, and Maxwell supposes similarly that the electromagnetic field is indeed such a system, yet in Maxwell's application to electromagnetism we find "mechanical" coordinates of very different kinds. One set of coordinates is "mechanical" in the ordinary sense; they are the physical conductors, the wires we can locate in space, and to which we can apply and measure Newtonian forces. Other coordinates, however, are "mechanical" only in a new, extended sense; these are the electrical coordinates: voltages, currents, and charge. Since in neither case do we see the material system to which we suppose they are all connected—the field—we may think of them throughout the following as bell ropes attached to an utterly invisible system. But they are bell ropes of two different categories.

There now follows a marvelous judgment-process, which must have few parallels outside of the realms of theology—for out of this abstract universal schema, we must select the actual terms that describe the electromagnetic world. The parallel I believe Maxwell keeps in mind is Newton's ceremony in "The System of the World," in which our gravitational universe is sorted out of the domain of all possible worlds that was developed in Book I of the *Principia*. In this sense, there might be said to be a *Principia* within the *Treatise*: chapter 5 of the *Treatise* is the new Book I of the *Principia*, while what we might call the Judgment of the Terms in the *Treatise* is the new "System of the World" of Newton's Book III.

Once Maxwell has assigned the generalized coordinates—the electrical measures of the currents, and the mechanical measures of the positions and motions of the wires that conduct them—it is possible to recognize certain of the corresponding coefficients in Lagrange's equations as

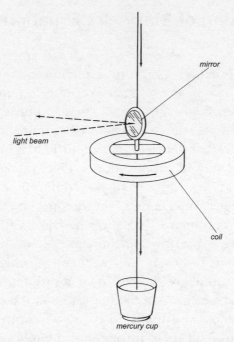

Figure 7. Experiment to detect momentum of a current in a wire. Arrows show direction of current flow: down the suspension wire, around the large coil, and out at the bottom through a mercury cup. [Redrawn from Tr ii/217.]

those that indeed do give rise to known phenomena. Other coefficients, on the other hand, seem to suggest phenomena that have never been seen. Perhaps those terms will have to be omitted as having no application to electromagnetism, but to make certain of this, Maxwell carries out certain novel empirical tests. In effect, the equations demand new, crucial experiments. They are "crucial" in the sense that, if they had succeeded, they would have become classic. Yielding negative results, on the other hand, a classic experiment may go unnoticed—though it is surely no less heroic in its conception, and no less significant in determining which of the possible worlds we live in, than the experiment of the crossroads that yields a positive "discovery."

One of Maxwell's crucial experiments that arises from the equations is diagramed in Figure 7, a drawing reproduced from

Figure 8. Maxwell's electrodynamic top, for the detection of a possible cross-coupling between electrokinetic and ordinary mechanical momenta.

the *Treatise*; had it succeeded, it would have revealed a momentum of the electric current as flowing in the wire itself. The absence of such an effect confirms for Maxwell his intuition that the real phenomenon to which we give the name "current" is not something that is "in" the wire. Another Lagrangian term suggests a cross-coupling between electrical and mechanical velocities—that is, a force arising from the combined effect of a mechanical velocity and an electric current. Maxwell's "electromagnetic top," depicted in Figure 8, was used to seek this phenomenon, again with a negative result. Maxwell was finding out systematically what world we are in.[27]

Having finished these preliminary judgments—having, in effect, chosen our universe—we now settle to finding the relationship between Lagrangian theory and the electromagnetic phenomena already well known. It will suffice, perhaps, if we watch this process in one representative case. Consider two coupled circuits, a primary and a secondary. Interruption of current in the primary gives rise to a sudden pulse of current in the secondary—stopping the first bell rope gives rise to a significant tug in the second: something is going on "upstairs"! To speak in Lagrangian terminology, the primary current was a generalized "velocity"—that is, the time rate of change of charge (not, we have learned, literally a velocity of flow of anything "in" the wire). Similarly, the "tug" on the secondary rope—the pulse of current revealed by the galvanometer in the secondary circuit—revealed the blocking of a momentum associated with the secondary: again, a generalized "momentum," and not one that corresponds to anything moving "in" the secondary wire. Maxwell speaks of this metaphoric entity of the secondary circuit as an "electrokinetic momentum," and the "tug" arising from its interruption as an "electromotive force." Note that these are strictly correct terms in the context of Lagrangian theory, however much the term "electromotive *force*" may distress modern students who recognize immediately that it is not a Newtonian force.

A metaphoric "mass," represented by an inertial coefficient in Lagrange's equations, relates this velocity in the primary to

---

[27] Maxwell's experiments have been repeated in various forms and have now yielded positive results, representing effects of the coupling of charge and mass in the electron. *See,* for example, S. J. Barnett, "A New Electron-inertia Effect ...," *Philosophical Magazine* 12 (1931), pp. 349ff.

momentum in the secondary. Where is this Lagrangian mass? Not in the circuits themselves, but in the field: somewhere (or everywhere?) in that "vacuum."

Maxwell now has the task of bringing all this back to Faraday. He has in view a way to draw out of Lagrange the *electrotonic state*, which, as we saw earlier, Faraday himself so long sought but never found. It was that "tug" in the secondary circuit that first led Faraday to believe that some tension was at that moment being released. But what Faraday saw as a *static* tension, Maxwell now perceives dynamically, as the electrokinetic momentum, and the tug not as the impulse from a released spring, but as the reaction of an interrupted flywheel. Faraday's insistent intuition of the importance of that "state" which he could never empirically verify appears now to Maxwell as an important leading clue—a sign that we are dealing, even in the quiescent magnetic field, with a powerful locus of kinetic energy. When we alter the field, we are tampering with the momentum of a very real mechanism—silent, intangible, and indeterminate in form and location as it may be. Its momentum is no less potent for its being known to us only metaphorically. Faraday had sensed its presence. This momentum, then, becomes for Maxwell the central concept of his theory. He finds ways to define an electrokinetic momentum vector at every point in the space surrounding a current; the summation of those vector components around a circuit is the total momentum of the whole current.

Yet we must remember that all of the phenomena we know, and all the values Maxwell is able to assign to the electromagnetic momenta corresponding to points in space, are only tugs on the bell ropes. The value of the field vector at a point is not the momentum *at* that point: it is only the momentum *referred* to that point—the pull on the bell rope, not the thing itself. We may arrive at a complete diagram of the configuration of the field—such as the one given, for example, in

Figure 6 (repeated).

Figure 6—but it remains only a "diagram." *What* and *where* the field is, we know only to the extent of this complete and quantitative metaphor.

What are known today as "Maxwell's equations" are partial differential equations expressing the relations at every point in the field among the quantities chosen as basic; they are the foundations of Maxwell's account of electromagnetism. Modern texts, taking these as the first principles of a mathematical *theory*, carry four such equations, and it is regarded as the elegance and success of what is called "Maxwell's theory" that more are not needed. Maxwell, on the other hand, wrote a dozen or so; he knew he could get by with fewer, but he thought that, if they suggested further ideas, the more equations he could conceive, the better. I think this suggests that he was not a theorist but an interpreter. Furthermore, watching their procession from Lagrange's equations, he was able to trace in them as they emerged their significance in terms of the dynamical theory—how they told of energy and momentum structures everywhere in the field. Today, I suspect, any physicist who paused to notice this would quickly set it aside as an unfortunate, though not very important, aberration.

The World of Light

We have followed one line of Maxwell's thought in his *Treatise* and have seen emerge from it first, the inversion of the world into the image of the electrostatic field, and then the dynamical system of the electromagnetic field. These two now prove to merge very neatly, the electrostatics entering the electromagnetic equations as a potential energy term. When that has happened, Maxwell finds that the connected system of Lagrange's equations has become a vibrant medium, capable of transmitting an electrical or magnetic disturbance with a very rapid, calculable velocity—calculable, that is, after some exacting measurements have been made.

The story of how those measurements came about is part of another reading of the *Treatise,* the *Treatise* in its context as a kind of Victoria Station in which people from the *Magnetischer Verein* and the Atlantic Cable enterprise are tromping through with their electrodynamometers and coils, while the Duke of Devonshire, as Chancellor of Cambridge University, is overturning the curriculum at Cambridge and fetching Maxwell from his retreat in Scotland to serve as the first professor of experimental physics and to build the Cavendish Laboratory. Out of that mix come the crucial measurements, the *Treatise* in its aspect of practical strategies and skill, which yield the electromagnetic theory of light: the demonstration, to convincing accuracy, that the velocity of propagation in the Lagrangian system is the very velocity of light itself. Light, then, is revealed as one form of electromagnetism. Then that metaphorical electromagnetic "matter" must fill the empty spaces of the cosmos wherever light can carry. This is the light-bearing aether, the discovery of which must have astounded Maxwell: the *electrotonic state* pervades and restores the cosmos, making of it one unified, communicating system.

This is perhaps a rather sad conclusion, for today every schoolchild knows that it turned out there wasn't any aether, and indeed it was rather naive of some old-fashioned gentlemen in the nineteenth century to imagine that there might have been one. We know now that science wasn't meant to deal with questions like that, but rather to write equations, make predictions, do experiments, and revise the results as necessary. In that way, we get on with what matters; the rest is only wasted time and blurred thinking.

I really do wonder whether we have quite understood all this. It would be another study to pursue the question further, but my sense is that between Hamiltonian relativity, which was probably in Maxwell's understanding as inclusive as Einstein's, and the critical and sensitive spirit he grew up with, Maxwell was philosophically ready for anything the twentieth century could bring on—I'm not altogether sure we are all, *post festum,* as ready as he was in anticipation. He was of course keenly aware of relativity questions; he had done a year-cycle spectrometer study looking for aether-drift effects, and one of his last letters was to D. P. Todd at the United States Naval Observatory proposing a study of the moons of Jupiter with this same purpose. In the *Treatise* itself, rather astoundingly, is to be found the theorem concerning moving conductors which can be understood as yielding the "Lorentz-FitzGerald contraction," the apparent "shrinking" of bodies as their speed approaches that of light.[28]

---

[28] This fantasy appears in connection with speculations about the comparison of the electromagnetic and the electrostatic systems of units, at [*Tr* ii/414].

Maxwell's convictions about relativity were deeper than any empirical detection of an aether could have shaken—he would have remained a relativist even if motion relative to an aether had been detected. His derivations of space and time from the phenomena of consciousness were by no means so absolute that he might not have been ready to cut a space-time continuum in various ways.

It is not important whether Maxwell foresaw this or that, or what he might have thought or seen had he lived to a riper age. The question is, I think, what he did see, and what we can find in the *Treatise,* not so much in terms of his having hit on right answers as having found ways to contain the right questions.

What I suppose I would like to ask is this: have we given up on Maxwell's effort to "wed Thought to Fact?" That does not mean going into the laboratory to test a prediction, or taking it out for confirmation on the Nevada flats. It means finding that relation between thought and a real world which gives meaning to what we think, or, put in a more antique style, preserves a relation to being. I think that is the difference between the "theoretical" and the "hermeneutic" approach to science. Maxwell, I have urged, was not a theorist (just count those equations); he was, or sought to be, a messenger. Only that, I think, can explain why Maxwell was not merely interested in Faraday, or primarily concerned with Faraday's results, but was passionately devoted to Faraday as a man. What could that mean? I believe that Maxwell saw in Faraday a fellow interpreter, one who understood the significance and profundity of a text which was sacred because it carried word of being, not in any fancy way, but in a simple, very human sense. Maybe all I am saying is that Maxwell seems to have seen science as a human activity, one that human beings could do, share, care for, and understand. I wonder if the schoolchildren all know, today, that that is a very old-fashioned idea.

Marx

Introduction to the Marx Essay

WITHIN WHAT WE CALL the sciences, we find the human spirit advancing freely and enthusiastically into the brilliant light of the 21st century, but in other matters it is hiding with Newton, Adam Smith, and Hobbes in the shadows of centuries long past. By some perversity of choice, we are modern in our sciences but archaic in our social thinking. Marx will show why this is so; the underlying causes are not laudable, but they are powerful.

Maxwell had envisioned human reason as being commensurate with whole systems, bringing to them an intelligent sense of larger human purpose. It is just such thinking that is needed if we are to deal with such global challenges as hunger, war, and the need for ecological sustainability—areas in which we are everywhere meeting defeat today. Our third essay, on Marx's *Capital*, will help us to understand why the world as a whole has so resolutely refused to join Maxwell's quiet, but potent, revolution.

Marx's *Capital*, I shall argue, is both an exemplar of standard science, and at the same time a work that lays out a new course and stakes out higher ground for science in the future. Just as we have seen Maxwell revolutionizing the concept of science inherited from Newton, so Marx is proposing a further transformation, to include society itself among the legitimate objects of the scientific mind.

Out of Marx's critical analysis emerges a new and vivid sense of what true human freedom would mean. His appraisal reveals an historic process, already far along in development, by which a new liberation of the human spirit is in preparation. Were it not for such insights as this, humanity might find itself sealed within a fate as inexorable as that in any ancient tragedy. But in ways I will outline, Marx's message is exactly the opposite: he sketches a brilliant image of the true meaning of human freedom and traces the way forward towards its attainment. To explore what that way entails is the aim of the essay which follows.

Toward a Reading of *Capital*

MARX'S *Capital* is not a book "about economics," any more than Plato's *Republic* is a book about politics. Plato uses the *polis* as a metaphor through which to investigate the human psyche and its practices, and Marx examines the structure of capitalism in the same spirit to ask us the most fundamental questions about ourselves and the world. In this commercial age, the economic metaphor is the more telling, but the question in a broad sense remains the same: who are we, and what is our practical relation to the human good? The *Republic* is strangely displaced when it is read as a text on "political science," and can really be approached in its integrity only when it is read as a letter from Socrates to us. In the same way, I want to suggest, *Capital* cannot be left to the economists; it may indeed interest them least of all, for they have other ways of answering the kind of question which specifically concerns them. But taken as a letter from Marx to us, and read as a challenge to our sense of values and the daily operations of our society, *Capital* remains a telling and disturbing work.

For Marx, "capital" is an enveloping system: no less than Plato, he is asking fundamental questions about our cities, our sciences, and our practice of the human virtues. If he is right, we are upside-down people, imagining ourselves in one world but living in another. In the belief that he sees something about us that we don't, I propose to consider what he has to say, and to this end I will ask the reader to embark with me on a brief dialectical tour of the three principal volumes of Marx's *Capital*.

Let us from the outset agree to dissociate Marx for present purposes from the Soviet Union and any other societies that have adopted his name and claimed his heritage. We need not here judge the question, which social projects have misused his name and which have read him well. At best, these are questions we could take up only after we had ourselves taken on the discipline of studying his work earnestly and as openly as possible. *Capital* is, I believe, a great book, enormously complex, and strangely subtle in ways the world has not often taken time to consider.

There is a special urgency in such an undertaking today, for *Capital* has ranked high for a century now on the world's bestseller list; it hardly needs to be argued that we must do everything in our power to maintain a substantive dialogue with the reading public of the world. Even if we ourselves saw no promise in Marx's text, or any relevance to our own lot, the fact that others throughout the world do find it so compelling would suggest that Marx is still speaking to the world's concerns; this fact alone would demand that we come to comprehend *Capital*, and read it, if we can, at least as well as others do.

Once we have agreed to address Marx and his book in their own terms, we are ready for a proposition which I would like to assert now, at the outset of our investigation. Marx, I believe, is *on our side.* By "our side," I presume to mean the side of democracy, and of respect for the worth and dignity of every individual person. The reader may judge Marx differently, but the difference is likely to be a conclusion that, in the end, Marx has not too little confidence in human capabilities but too much—that he would grant us too much freedom, too much responsibility, that he credits too highly our powers of using our human skills to plan together rationally for the achievement of our common human goals—in short, that Marx is too democratic. In any case, I suggest that Marx is challenging us in *Capital* to go further—much further—along the way of democracy, down our own road, than we have gone ourselves.

Marx respects the central documents of the American tradition; these he sees as claiming and formally granting the equal rights of every individual. All that Marx believes and proposes, he bases on the conviction that these rights have been won and must be defended and preserved. But as *formal* rights, they only launch the project of human freedom. The formal freedoms won by the American Revolution have yet to be implemented in a society that makes the corresponding real freedom a fact in the life of every member. A candid view of contemporary America can only remind us how far we have to go to achieve such a reality—and we are, too slowly, learning that we cannot in the modern world think of such matters in terms of one nation alone.

Marx may have more faith in us than we have in ourselves. He believes that it lies within the powers of modern industrial society—even *his* "modern" society of a hundred years ago—to make these freedoms *real.* As we shall see, this for him takes the form, not primarily of a redistribution of the world's present wealth, but of a transformation of society itself into a more rational human community, and the reorganization of the productive powers of modern technology so as to make real individual freedom and opportunity practicable. These are bold proposals, but as ideas they are rather simple, and they might be taken as nothing more, or less, than proposals for the implementation of the objectives of our own democracy—were it not that the image normally associated with Marx is totalitarian, so that the reader cannot believe they are what Marx really means. I propose that we listen to Marx himself, and see what he has to say. [1]

---

[1] In *Capital,* Marx envisions "a higher form of society, a society in which the full and free development of every individual forms the ruling principle." (Vol. I, chap. XXIV, sect. 3; p. 555) In general, the discussion in this essay will follow the order of the argument in *Capital* itself, and I will try to let the reader know, in footnotes to the headings of this study, where we are in Marx's text. In this way, the interested reader might wish to use this study as a partial guide to reading selected sections of *Capital. Capital* is usually thought of as consisting of three volumes, of which only the best-known, the first, was seen through publication by Marx himself; Volumes II and III were edited and published by Friedrich Engels. (*continued overleaf*)

I think it is especially interesting that he writes, not as a Utopian or an idealist, but, he claims, as a scientist—a *dialectical* scientist. With this claim, he is challenging our concept of science itself; in effect, he finds that criticism of our ordering of political and economic society entails criticism of our practice of what we call "science" as well. A society and its sciences are, it seems, images of one another. We shall have to see, as we go along, how this might be understood. But for Marx it does seem to mean at least that he for his part understands *Capital* to be a work of science, and not just political rhetoric or speculation. We may thus have to reconsider our own concept of knowing, and of ourselves as knowers.

Marx may be right or he may be wrong (and we, in our lifetimes, may never know which). But his propositions stand in either case as fair challenges to our understanding. I hope that this present essay will help to open again the reading of *Capital,* which in our time of reaction is so difficult to undertake with that readiness to listen and, perhaps, to learn, that every great author must presuppose in order to be heard at all.

Formal and Dialectical Theory: the Plan of this Study

Marx is indeed confronted with a difficult concatenation of tasks. Both he, as author, and we, his readers, are totally embedded in an intricately woven historical era, the age of capitalism. True, he has arrived at it a little earlier than we,

---

In this essay, references are made to Volume I in the translation of Samuel Moore and Edward Aveling. Reference will be made to the last two volumes in another edition, translated by David Fernbach. There is a "Volume 4" of *Capital* (in three volumes!), edited by Karl Kautsky under the title *Theories of Surplas Value*; it is of special interest to readers of the *Great Books,* since it includes the critique which results from Marx's intensive reading of Adam Smith.

and assuredly there have been major shifts and changes from his time to ours, but essentially we share with him what Hegel would call the "spirit" of a common experience. The institutions we know, the language we speak, the social relations we have grown up to accept and work with as the terms and conditions of our lives, belong to the system of thought and practice which is capitalism: private property and its law, the modern city, the modern state, buying and selling, investment, profit and interest, and that kind of working-for-pay which we take for granted and call "having a job," "earning a living," or "pursuing a career." They are the materials of our lives (and it is just by virtue of his attention to the importance of the detail of these environing conditions of our lives that Marx is—appropriately—called a "materialist"). Yet though they are the material of our lives, we know, when we pause to reflect upon it, that they are products of history—they once were not—and they are, in principle, transitory. Capitalism is a brilliant and seemingly ever-more-total world; but it is a world that is a product of time and is thus in transience.

Thus, from a point *within* capitalism Marx proposes to develop a theory of capitalism. Already this is strange ground, for one expects to meet theories of objects that are permanent and about which we can be "objective"—and that are, as such, appropriate concerns of strict science. Marx will construct, instead, a formal theory of a transient object, and one with which we are totally and subjectively involved. Worse, though, the theory will be critical: that is, it will present its object, not only as transient, but as perverse—as inherently contradictory, and as an alienation of our humanity. And finally, if Marx is right, this science of a transient and perverse subject matter must be constructed from within that domain itself, using the language of an alienated dictionary and the experience and thought-structures of alienated minds, both his and ours. I think we can rightly characterize this as a dialectical enterprise, in the sense that it is fundamentally critical: it questions the very ground it stands on. It probes its own terms and our cherished concepts, and it reaches for higher ground, looking to the long scope of history, and speaking already in a language that must borrow problematically from what it conceives to be the human future.

I think we can conceive the work of *Capital*, then, in two distinct aspects. The first is that of the formal theory of capitalism, a science of this transient object, which will undertake to depict the operation of capitalism as a system, accurately and objectively, much as Newton depicted the motions of the heavens. For Marx, this will be, in effect, the science of a passing object, the truth of an error. It will view capitalism objectively, taking it provisionally in the way in which it presents itself in our time.

The second aspect is the critical theory, which undertakes to expose what Marx sees as the error of capitalism and locates it in a larger horizon of the past from which it came and, above all, in its potential historic destiny. Here the terms that are accepted by capitalism are criticized and their larger meanings revealed; the illusions generated by capitalism are identified and traced to their sources, and above all, perhaps, the workings of inner contradictions explored, by which capitalism is seen to be generating, within itself, its own antithesis. This we may call the dialectical theory of capitalism.

I suggest we can think of these two works of *Capital* as nested, one within the other—the *formal* theory, as a limited enterprise, contained within the encompassing *dialectical* theory. We might speak of the formal theory as "framed" within the dialectical. I do not mean to say that Marx addresses the two theories sequentially or in separation; elements of both are interwoven throughout the book. But they are in principle distinct, and we will do well here to take them up, as nearly as possible, one at a time. I propose, then, that we first address the formal theory and then consider the larger dialectical complication. Perhaps it is just as well that in practice things cannot be quite so neat, but we may take this plan as our guiding thread.

The Cell Form of Capitalism: The Commodity[2]

Indeed, at the very outset we meet a difficulty. Before the formal theory can unfold, we must establish a starting point, and this is not so easy. "Every beginning," Marx says, "is difficult." (*Capital*, Vol. I, p. 18) Marx must find the right point for the initiation of his own theory of capitalism for he cannot assume that everyone will start at the same place. His beginning point, he decides, must lie in the concept of the commodity and the commodity proves to be a paradoxical and elusive entity. To locate it, and thus to found the formal science, is already a dialectical task: that is, Marx must focus on a common concept, accepted within capitalism as altogether obvious, and examine it afresh with dialectical rigor. Only then will he be able to proceed securely with the formal science which builds on it. This dialectical examination he speaks of as an analytic procedure, likening it to microscopy. As the microscope reveals the cell as the common unit of living things, unseen by the unaided eye, so Marx's analysis identifies the commodity as the universal principle of the system of capitalism (Vol. I, p. 18).

What happens to an object, Marx asks, when it becomes a "commodity"? Like a Socratic question about virtue, justice, or love, this one begins innocently enough but expands in significance as we pursue it. The beginning is with the distinction, already familiar in other authors before Marx, between "use-value" and "exchange value." The objects we employ in our lives have value for us because they serve our purposes: many are simply "useful" in that they meet necessity, providing food, shelter, or clothing; others we prize for their beauty and delight. All, as serving our human ends, have "use-value." The term is hardly adequate. Its referent is all the world of *quality*— of appetite, color, warmth, friendship, intelligence. Whatever

---

[2] Marx discusses the "commodity" in Part One of Volume I of *Capital* (Vol I, chap. I; pp. 43–87).

we value and strive for belongs to this realm. It is important to reflect upon it here, for we are about to leave it behind.

Within our capitalist era, virtually every object we lay eyes upon has its price: most of the things we see about us have been objects of exchange and could, on some secondhand market, fetch a price again. Some have their price tags still on them: books tend these days to have selling prices printed on their dust jackets. Even a human life, it might seem, has its price: the courts are daily concerned to evaluate disabilities and deaths in terms of dollars and cents.

Things that are on the one hand needed, prized, or loved by human beings, but are at the same time made, bought, sold, and evaluated in money terms as objects of exchange, are commodities. In their qualitative role, serving human ends, they have use-value; yet as traded, bought, sold, and exchanged, they are reduced to equivalent *quantities,* and have exchange value. How does the world of quality thus turn into a world of quantity—and what does it mean for us, that it should do so?

Marx finds ingenious ways to examine this question in Chapter 1 of *Capital*; he weighs and scrutinizes what he calls the "forms" of the commodity and commodity exchange, dwelling first on one aspect, then another, and in the process the reader learns not only much about the commodity, but about Marx as well, and the meaning for him of "analysis." We meet a powerful, relentlessly critical mind, "analytic" not in the modern, algebraic sense but in some reconstitution of the Aristotelian patience with the particular. Out of it comes a steadily growing sense of the pervasiveness and significance of the "commodity" as founding idea of capitalist society: the transformation of quality into quantity, of all things into numbers which rate them in exchange.

I want to suggest that this first chapter of *Capital* sounds the note of Marx's whole work, and that his style seems to me to have something of the ancient about it, rejecting contemporary analysis and our modern, easy use of symbols, and insisting instead on finding the *content* of terms and transactions. He will take, for example, an ironic instance of an exchange—linen for Bibles—and work it out, back and forth, in arithmetic concreteness (Vol. I, chap. III, sect. 2; p. 107). Modern readers may find themselves longing for an algebraic equation, or a graph or two, to summarize in a schema the relations which these details are tracing, but I think hardly in the thousands of pages of *Capital* does Marx ever really gratify that wish. He is on a different track, humanly and intellectually. And I think we can see a clue to this almost stylistic issue in the concept of the commodity, for what Marx is discovering in this "Commodities" chapter is the universal substitution of market magnitude for the distinctive qualities of those things in our lives which meet human needs, or answer human desires. In his own style, Marx will not follow that lead.

That substitution of quantity for quality, the manipulable symbol for a lost content, is exactly the questionable algebraic magic of the "Cartesian Revolution." That brilliant but disturbing exchange which is accomplished in the drama of the *Discourse on Method*, Marx in effect refuses to accept at face value. There, Descartes once and for all, on our behalf, doubts away the old world of quality, color, sound, and the human good and returns from the depth of his isolation with a new world of mathematical magnitude and formulable relation (*Discourse on Method*, II–IV). That is strikingly close to the universal transformation of use-value into exchange value of which Marx speaks, and Marx insists on reminding us of what we have lost. In his own manner and thought he is taking his stand outside that transformation; he will not reason as a Cartesian but stands resolutely and comfortably with the ancients in this, and perhaps with those of a future which will have passed through and beyond the Cartesian blackout.[3]

---

[3] It may be helpful to point out, at the outset, that Marx was in certain ways in fact a "classicist." His doctoral dissertation was on Democritus: my own sense of this is that he was especially interested in the concept of free will or spontaneity, represented in Democritus's theory by the "swerve" of the atoms. (Lucretius includes this in *De rerum natura* [II, 216].) Marx knew Greek well, and his (*continued overleaf*)

The commodity is not, of course, in itself a creation of modern capitalism. Marx catches Aristotle in reflection on this same mystery of the transformation of quality into quantity, in the markets of Athens. By what common measure, Aristotle asks in the *Nichomachean Ethics* (1133b 25), can five beds be equated to one house? Substantively, they are utterly unlike, yet they exchange in a specific numerical relation. Marx reports that Aristotle finds no adequate answer, and Marx in turn finds Aristotle's bewilderment appropriate and significant: the full, illogical logic of the commodity is the work of history, and in Athens there was at root, he says, no rational answer to Aristotle's conundrum.[4]

Only with capitalism has the commodity fought through to a completely coherent system of reckoning. We shall see in a moment Marx's solution to Aristotle's riddle, the foundation that Marx finds in capitalism for the ratios of universal exchange. But what is most striking in this passage is the evident satisfaction with which Marx accepts Aristotle as a colleague in the contemplation of the question: for both, the commodity is a dialectical problem, not an acceptable fact.

To help us in recognizing the profundity of the world change which the unbounded logic of the commodity has

---

biographers report that he read Aeschylus regularly, throughout his life. "According to Lafargue Marx read Aeschylus in the original Greek text at least once a year." (Franz Mehring, *Karl Marx*, p. 503); Marx listed Aeschylus as one of his three "favorite poets"—the others, Goethe and Shakespeare.

[4] Aristotle says, in the *Nichomachean Ethics*, "Now in truth it is impossible that things differing so much should become commensurate…." (1133b 19) Aristotle sees need ("demand") bridging the gap in practice, but what interests Marx is that Aristotle recognizes the lack of a common measure in *principle*. What was lacking, Marx says, was the concept of *human* equality, as the needed common unit of measure.

brought with it, Marx turns his inquiring irony to the contemplation of a chair. Its analysis, he says, "brings out that it is a very strange thing, abounding in metaphysical subtleties and theological niceties":

> So far as it is a value in use, there is nothing mysterious about it, whether we consider it from the point of view that by its properties it is capable of satisfying human wants, or from the point that those properties are the product of human labour. (Vol. I, chap. I, sect. 4; p. 71)

It begins as a thing of wood, "an ordinary sensuous thing." Yet:

> …so soon as it steps forth as a commodity, it is changed into something transcendent. It not only stands with its feet on the ground, but, in relation to all other commodities, it stands on its head, and evolves out of its wooden brain grotesque ideas, far more wonderful than "table-turning" ever was. (Ibid.)

Where does this mysterious "value," which the chair possesses as a commodity, come from? It is nowhere visible to the eye, nor is it an object of touch; it is hardly related to use-value, since the items in the world that are most vital to us—the air we breathe, for example—often bear no exchange value at all. Marx's comic, Till Eulenspiegel chair figures the world turned upside down: it is a comic image of a tragic transformation, in which each object in our world converts itself from quality to quantity, from human service to the overriding demands of a universal, measuring market. If the wealth of our society indeed has taken the form of "an immense collection of commodities," then wealth has, *per se*, become stark *amount*, bearing, Marx says, "not an atom of use-value." We are at the threshold of a system of alienation of use-value and the human, the insight that for Marx is the key to understanding the entire world of capital.

Labor Value [5]

We see that if Marx is to set forth a consistent theory of capitalism, he must first resolve Aristotle's conundrum and specify that common measure that will bring all things, as commodities, into quantitative relation.

Although I see no evidence whatever that Marx himself had this analogy in view, I would like to suggest another measuring task at the foundation of a formal science, which I think bears a striking and instructive likeness to the problem Marx faces at this point. I suggest that Marx is very much in the position at which Newton stood at the threshold of the *Principia*. Like Newton, Marx must choose a measuring unit that is well founded, and that will make a world intelligible. For Newton, this was the world of the planetary motions. For Marx, it is the world of capital.

At the outset of the *Principia*, Newton defines the quantity which he calls "mass"; this is a bold move, and one on which the integrity of his science entirely depends (Book I, Def. 1). Correspondingly, Marx defines the unit of exchange value. There was nothing obvious about Newton's choice: indeed, "mass" is a concept remote from the familiar measuring

---

[5] The fundamental principles of Marx's labor theory of value are set out in the opening chapter of Part One, on commodities, to which we have already referred in general (Vol. I, chap. I). The labor theory of value is discussed, with reference to John Locke and Adam Smith, in the *Syntopicon* article on LABOR, to which the reader may wish to refer (*GBWW*, Vol. 2, pp. 926–27, 936). Marx certainly read with great interest such passages in Smith as this: "Labour, therefore, it appears evidently, is the only universal as well as the only accurate measure of value...." (*Wealth of Nations*, Book I, chap. 5)

processes in mechanics. We do not compare bodies by determining their masses; we put them on balances and determine
their relative weights. Masses obey laws—Newton's "Laws of
Motion" at the outset of the *Principia*—that are counterintuitive
and that were, until recently, virtually unobserved in practice;
according to Newton's first law, a body once set in motion will
move with that same speed in a right line forever, whereas we
all know that a terrestrial body will in fact shortly come to a
stop (Book I, Law I).

Marx's definition of exchange value is almost equally
difficult to grasp, and modern economists disdain it as unnecessary and contrived. I propose, however, that like Newton's
"mass," Marx's definition of value is the cornerstone of a strict
and powerful theory. The reason economists today do not
appreciate this is, very possibly, that they have a different idea
in mind of what a theory should be, and they do not demand
an economics, as Marx did, that has the kind of rigor and
coherence that distinguishes the *Principia*'s account of the
heavens.

Marx asserts that the exchange value of a commodity is
determined by the amount of human labor it contains. In saying this, he is repeating an opinion that was widely shared by
such predecessors as Adam Smith and Benjamin Franklin. The
difference is only that Marx carries the principle through with
a precision and consistency that no one else had attempted.
Marx states:

> That which determines the magnitude of the value of
> any article is the amount of labour socially necessary
> ... for its production.

Further:

> Commodities, therefore, in which equal quantities of
> labour are embodied, or which can be produced in
> the same time, have the same value. The value of one

commodity is to the value of any other, as the labour time necessary for the production of the one is to that necessary for the production of the other.

Finally as exchange values:

> …all commodities are only definite masses of *congealed labour time.* (Vol. I, chap. I, sect. 1; p. 40. Italics added.)

Clarification of this concept to the point of scientific strictness presents a series of problems, which Marx carefully and systematically resolves. The first challenge is to clarify the notion of "socially necessary labour time." As Marx points out, if we simply define value as "labour time," it would follow that the product of a lazy or unskilled workman would be worth more than the same object produced by one who was energetic or skilled. It is for this reason that the unit of measure of labor time must represent a social average:

> The labour time socially necessary is that required to produce an article under the normal conditions of production, and with the average degree of skill and intensity prevalent at the time. (Vol. I, chap. I, sect. 1; p. 39)

Indeed, not only must the labor unit represent the social average for any given industry, with socially average skills and the level of development of machinery that customarily determines the productivity of labor in that branch of production; we must now speak of a unit of labor that is common among the whole variety of industries whose products enter into exchange in the total market system of an economic society. As a commodity, the object of labor becomes a certain quantity of exchange value, and as such, it has "abstracted," Marx says, "from the material constituents and forms which make it a use-value." It is:

no longer a table, a house, yarn, or any other useful
thing. Its existence as a material thing is put out of
sight. Neither can it any longer be regarded as the
product of the labour of the joiner, the mason, the
spinner, or of any other definite kind of productive
labour.... There is nothing left but what is common to
them all; all are reduced to one and the same sort of
labour, human labour in the abstract. (Vol. I, chap. I,
sect. 1; p. 38)

The commodities all tell us, Marx says (and we may imagine
ourselves gazing in bewilderment down the endless corridors
of the supermarket or the discount store as we read these
words) "that human labour power has been expended in their
production, that human labour is embodied in them. When
looked at as *crystals of this social substance,* common to them all,
they are—values." (Ibid.; italics added.)

Newton, as we have seen, similarly founded his system on
an act of radical and total abstraction—abstraction from every
quality of physical substance but that invisible one, *mass*: not
mass of copper, or of air, or of flesh, but of undifferentiated
matter, a property exactly as unimaginable as Marx's "social
substance." In either case, only such an act of penetrating
abstraction would reach through to a unit sufficiently universal
to weave together a single, coherent, and inclusive world.

I mentioned in passing that modern economists generally
disdain this "labor theory of value" as unnecessary to their
science, and I suggested that this is because they have a different idea of "science" itself. Let me suggest further what that
alternative view of science might be, for I think that in this way
we may locate more accurately what Marx is attempting, as well
as the quite different, though perhaps equally legitimate, alternative aims of contemporary economics.

Two paradigms of science stand prominently before us in
our Western tradition: that of Newton, to which I have been
likening Marx's formal theory, and another, very different, that

is represented first and perhaps best of all by Ptolemy in his
Almagest.[6]

The history of astronomy follows the trail of their divergent
conceptions and purposes. Ptolemy's aim is, he says, to "save
the appearances." He develops impressive, far-reaching mathematical methods (first cousin to Fourier analysis, one of the
principal tools of modern mathematical physics) to achieve
this purpose, but through it all, he regards the planets for his
mathematical purposes merely as points of light; projected
onto the celestial sphere, they are nowhere, and if they are
made of anything, it does not concern his science of astronomy
to consider that. His purpose is strictly and exactly limited: to
find the mathematics of the observed motions of the heavenly
bodies as geometric figures. Where and what they are, and why,
concerns physics or theology, and these are neither of them
the science of the *Almagest.*

It is such a paradigm that I believe the modern economist,
like the modern quantum physicist, in effect looks to. To find a
predictive mathematics that worked would be all that the contemporary scientist in either area would demand. Not so with
Newton, or with Marx: each identifies his science with a right
and penetrating determination of the substance he has taken
as his object. Ptolemy, and our contemporary economist and
quantum theorist, are quite reasonably content to *save* their
appearances: Marx and Newton are determined to *penetrate* the
appearances and save the underlying substance. Marx, I want
to claim, is a "classicist," not only in his unwavering attachment
to Aristotle and the tragedians, but in his perhaps intuitive

---

[6] Ptolemy confronts a crisis when he finds that two different hypotheses will account for the same phenomena, namely, the epicycle and
the eccentric. His comment is, "…all the appearances can be cared
for interchangeably according to either hypothesis." That being
the case, he concludes it is reasonable to opt for the simpler one
(*Almagest*, III.3). On the idea of SCIENCE, *see* the *Syntopicon* (*GBWW*,
Vol. 3, pp. 682–705). This article points out that Marx in effect takes
England as his laboratory (ibid., p. 686).

attachment to the goals of what we now call "classical physics." Marx, like Newton, is most concerned with the right under- standing of the substance on which the science is founded. It is in this sense that he said, "every beginning is difficult."

It is often remarked that *Capital* advances through descend- ing levels of abstraction, and Marx himself makes clear that the beginning has been accomplished through an act of extreme abstraction from experience. Certainly the concept of "undif- ferentiated homogeneous labour time" is utterly remote from any particular, concrete labor process, just as a chair regarded as a "commodity" is utterly abstracted from anything one might sit on. In this sense, *Capital* begins at a pinnacle of abstraction, and the course of development of the work will be through a series of descents to successively lower levels until finally we find ourselves back, at the close of the formal theory, on the plane of the phenomena themselves—the daily phenomena, that is, of capitalism: actual market prices, profits, interest rates, and rents.

We must be on our guard, however, for there is another sense in which in Marx's case the concepts of "abstraction" and "descent" are complicated and even, perhaps, reversed. Marx is, after all, in his own view building a science of an alienated, illusory realm: he is setting out, as we said, to tell the truth about an error. The error begins with the commodity, in which use-value, about which we care, is supplanted by exchange value, which impersonally governs the motions of capitalism in its stead. Marx's formal science is the theory of exchange value, the false facade of use-value. Only with the dialectical theory, which I have envisioned as containing and "framing" this formal theory, will truth come into focus in its own right. The world whose theory we are now building, the world of commodities, even when these take forms of the most vivid phenomena in "concrete" detail—that world of market prices and the daily Dow Jones quote—this is itself an abstract world.

What is ultimately in fact substantive is the other face of the commodity, which capital in all its transactions leaves behind:

use-value, human interest and purpose, and the concerns of the original "economics." Thus the further we go in constructing the edifice of capital, which is of course Marx's purpose in the contained, formal theory, the more complete is our severance from all relation to the genuinely human world, the only world that is ultimately "concrete." In this sense, by building a world of abstraction, we will only be sealing and confirming more completely our alienation from the concrete. We will "descend," then, to the detailed phenomena of an alienating, illusory, and abstract world, though nonetheless the very one in which we live, and with which we are most familiar. When we return to it under Marx's guidance, we will find that it looks very different; Marx will bring us back in a dialectically controlled reentry—with changed minds, if his enterprise has succeeded.

We must agree with Marx that our starting point, "socially necessary homogeneous labour time," is highly abstract, and difficult or impossible to visualize. How, then, can this elusive notion serve as a firm foundation for an applicable theory? If we cannot even visualize it, how can we expect to use it as our platinum-iridium measuring rod for the determination of exchange value? Efforts to characterize it more fully hardly make things any easier to grasp. Marx suggests, for example, as he often will do as the work progresses, that we think in terms of society as a whole. Will this help? He invites us to consider "the total labour power of society" as "one homogeneous mass of human labour power, composed though it be of innumerable individual units." Pursuing this suggestion, he asserts:

> Each of these units is the same as any other, so far as it has the character of the average labour power of society.... (Vol. I, chap. I, sect. 1; p. 39)

This may give some hint of a way to consider the question, but it surely does not resolve it. Do we have, then, a firm foundation for a science, or do we not?

Fortunately for Marx's science, the theory in no way depends on our ability to produce an exemplar—or a mental image—of this socially average labor power. The exact and scrupulous determination of the labor value of commodities is a practical measuring process which is in fact going on in the markets of capitalism at all times. Marx has precisely characterized a process, a relentless equipoise of product against product, by which the universally interconnected markets of capitalism are always appraising precisely the quantity we are interested in. As bargains are struck in the myriad daily purchases and sales of the commodity markets, what are in effect *measurements* are made which are more exacting, perhaps, than those of a bureau of standards. Overpriced products are discounted and must come down in their prices; those that are underpriced are quickly bought up, and their places are filled by others that are more realistically evaluated. The free market is the laboratory of capitalism. Marx has not dreamed up the definition of value: he has simply penetrated the secret of a historical fact. The weighing, evaluating, and appraising of commodities in terms of that single homogeneous unit is the underlying *fact* of the universal system of markets in capitalism; it is what is going on around us, every day. All products are in fact being brought into ratios with one another by the market process. Such a thoroughgoing technology of universal exchange did not exist in Athens, and Aristotle was rightly baffled by the problem of relating five beds to a house. History has solved Aristotle's problem, and Marx has simply interpreted that solution.

Capital and Wage Labor[7]

Marx is taking us on a revelatory tour of the world in which we live, whose name is "capitalism"—the reign of capital. What, then, is "capital"? Conventional wisdom answers in terms of

[7] Vol. I, chap. IV; page 146.

things: a stock of goods, an investment in factories, processes, and tools—the means of production. Marx turns our attention instead to the form of organization of our society, the social relations under which we live. These are, of course, characterized by the institution of private property—more specifically, private property in the means of production. Yet this does not quite suffice to define our own era: means of production were owned before capitalism came to birth. Nor is it sufficient to speak of a class division, for there were classes of the rich and the poor, the few and the many, in ancient Greece. Only the combination of these two criteria catches the essence of our historical situation: exclusion of the worker from ownership of the tools with which he works, so that private property in the means of production is concentrated in one class, while the other, that of the workers, is obliged to come to these owners as wage labor in order to work at all. It is really, then, the *concentration* of ownership of the means of production—made possible by the universality of that supremely accumulable commodity, money, with its surrogate, credit—that gives rise to the specific social relations constituting our economic world.

It is often suggested that Marx writes of another era, that his analysis may have been appropriate in England of the nineteenth century, but that we have passed beyond the social conditions he describes. I think this perception, on the part of those who are indeed quite understandably struck by the conditions of long hours, undernourishment, and child labor which Marx describes, misses the central point of his definition of "capital." We are today a class society, and we live under the relations of capital as he describes them insofar as there are among us those who must seek employment by others in order to carry out our own productive work. This is true whether that employer is a person or an institution, and whatever form the "work" may take—wage labor or salaried employment, production or service. For "labor," at least to a first approximation, we may read "employment"; and "class" is still the central fact of capitalism in the sense that most of us must seek employment from others, in order to work and live. This concentration of ownership of the means of production has of course become much more extreme in our time, the age of vast corporations and multinationals, than it was when Marx wrote. And while we may be a mobile society, with opportunity for all, as our mythology reiterates, the fact of concentration of ownership remains impressive.[8]

When we enter into an employment contract, we transfer to another a certain right—the right to command a specified amount of our labor time, and to own the product of our work. This has nothing to do with the modality of the work—whether the agreement is friendly or not, whether we expect to enjoy the tasks we are given, whether in fact it may seem to us the very "opportunity," as we say, for which we have been waiting. It is a legal contract (written or not), the very foundation of our social order: our work, and our work product, by right belong to another. This means, as Marx points out, then, that our labor power has become a commodity, entering a market as other commodities do (the "job market," we do not hesitate to call it), which appraises the items offered for sale and strikes bargains with at least as much rigor and strategic care as in the traffic in hogs, or wheat futures.

The fundamental insight here is that all markets are inherently coupled in a single system, since money can flow freely from one to another. If on a given day we are a better bargain

---

[8] Many studies, dating from those of the Temporary National Economic Committee in the 1930s, have shown this; recent studies seem only to confirm the old findings. One analysis, for example, found that in the realm of personal ownership of corporate stock— close to our present concern—83 percent was in the hands of the top 5 percent of the population (cited in Richard Parker, *The Myth of the Middle Class* [New York: Liveright, 1972], p. 212). The subject, admittedly, is one which invites bottomless discussion on the part of professional economists and sociologists, but data of this kind abound, which strongly indicate the unequal division of wealth, power, and opportunity in our society.

than hogs are, money will be siphoned out of hogs and put into us. Capitalism entails the universalization of the commodity-relation and, by virtue of the private ownership of the means of production, comes to include, among the other commodities, man himself.

The universal role of money, which has made unbounded accumulation possible and thus opened the way to capitalism, has likewise restructured the practice of commodity exchange in a fundamental way. Marx facilitates our thinking in these matters by means of a striking formula, which he has devised to represent in one linear chain of symbols the flow of an economic process. Commodity exchange is represented in this way:

$$C—M—C'$$

A commodity, C, is brought to market and exchanged for money, M, while that money in turn is exchanged for a second commodity, C', for which the original commodity owner has greater need. Here, a human need has been satisfied. (Vol. I, chap. III, sect. 2a; p. 106.)

It would represent a total change in social relationships if we were to write the same formula from a different point of view, starting now with money rather than with a commodity:

$$M—C—M'$$

Let us call the one who enters the market carrying with him, not a product, but simply money, a "capitalist." Unlike the shoemaker, who comes to market to sell a commodity of which he has an excess in order to buy something for which he has a need, the capitalist comes to market with money. He buys in order to sell: that is, both the beginning and the end of his cycle will be money itself—initially M, and finally M'. There is, of course, something absurd on the face of this transaction—if goods trade at their values, as Marx has presupposed earlier, the beginning and end of the process consist of the same thing, and nothing at all has been achieved. Marx puts this puzzle in stark terms. The capitalist, he says:

> ...must be so lucky as to find, within the sphere of circulation, in the market, a commodity whose use-value possesses the peculiar property of being a source of value, whose actual consumption, therefore, is itself an embodiment of labour, and, consequently, a creation of value. (Vol. I, chap. VI; p. 167)

Only in that case can M' be greater than M, and the capitalist's activity make sense. The quotation above concludes:

> The possessor of money does find on the market such a special commodity in capacity for labour, or labour power. (Ibid.)

To explain this claim, Marx makes a fundamental distinction, the roots of which I think must be found in his frequent mentor, Aristotle. The commodity in question is "labour power," not "labour." He buys the *potential* for labor; Aristotle's term would be *dynamis*, "potentiality." The use-value of this potential is labor-in-act, *actual* labor: this corresponds to Aristotle's term, *energeia*, "activity." The difference, Marx says, speaking with a voice that sounds very much like Aristotle's *Physics* at this point, corresponds to the difference between the *faculty* of digestion and the *act* of digestion.[9]

Now, the economic significance of this distinction in the grammar of economics is immense: for if the commodity is labor power, it will be priced like any other commodity at the cost of its production. The value it produces, on the other hand—its

---

[9] I am thinking in a general way of the fundamental account of motion in Aristotle's *Physics,* Book III. Marx's understanding of our freedom as being experienced in *activity* ("self-activity" is a term he uses often) is suggestive of Aristotle's emphasis on *act* (*energeia*). (200b 7ff.)

use-value as labor-in-act—will be all the value it can yield in
the course of a working day. Between the cost of production of
labor power and the value that labor power can produce when
put to work lies the difference between M and M' in the formula for capitalism—and hence, the fundamental principle of
profit and "return on investment." Labor produces more value
than it costs and, under the social relations we are describing,
that added value belongs not to the laborer but to the owner of
the means of production, the employer of the laborer.

Marx gives to this expansion of value the name "surplus
value," and we may rewrite the formula for capitalism to give
this explicit recognition:

$$M—C—C'—M'.$$

Here C includes the labor power, which is the magic commodity the capitalist purchases, while C' represents the greater
value of the product which that labor produces in act—the
difference $(C'-C)$, or equally $(M'-M)$, is the "surplus value"
created in the process. This formula may be taken as the law
of motion of capitalism, since the prospect of this expansion
becomes the driving force of all investment. It is a law that
seems to me strikingly analogous in principle to Newton's
Second Law of Motion of the heavens:

$$M—C—C'—M'$$

(Marx)

$$F = ma$$

(Newton).

Each is universal in its own domain. Newton's law sees a
physical force giving rise to an increment of velocity in a given
time; that of Marx speaks of a motivational force giving rise
to an increment of money in a cycle of production. There is
no acceleration in the heavens without what Newton calls a

"motive" force; there is similarly no investment under capitalism without the prospect of gain, of which the employment of living labor is ultimately the only source. We are on the track of a dynamics of economic motion very much in the Newtonian, classical sense.

In the commodity-market equation $C—M—C'$, one quality was exchanged for another; the market process had a natural purpose and a limit with the completion of the transaction, in which something unneeded had been exchanged for something that was wanted. It made human sense. In $M—C—C'—M'$, which can be contracted from the point of view of the investor to $M—M'$, the alpha and omega are mere quantities. Money, which entered in the era of commodity exchange as a medium of exchange, has now become the goal. Quantity, rather than quality, rules the process. One quantity as such, however, is like every other quantity: the process that once had a natural end and made sense is now inherently infinite, endless. Marx observes, of M and M':

> … both have the same mission to approach, by quantitative increase, as near as possible to absolute wealth…. The circulation of capital has therefore no limits. (Vol. I, chap. IV; pp. 151–152)

It is interesting that Marx here once again permits himself a long reference to Aristotle, in which he reviews the latter's thinking about this question of the conversion of quality to quantity in economics. Aristotle sees a degradation of "economics," from that art which aims at the provision of human goods (the Greek word, *oikonomike*, refers to the art of household management), to a mere art of moneymaking, which he names "chrematistic," after the Greek word *chremata*, "money":

> … in the case of Chrematistic, circulation is the source of riches. And it appears to revolve about money, for money is the beginning and end of this kind of

exchange. Therefore also riches, such as Chrematistic strives for, are unlimited. Just as every art that is not a means to an end, but an end in itself, has no limit to its aims while those arts that pursue means to an end are not boundless, since the goal itself imposes a limit upon them, so with Chrematistic, there are no bounds to its aims, these aims being absolute wealth…. By confounding these two forms, … some people have been led to look upon the preservation and increase of money *ad infinitum* as the end and aim of Œconomic. (Vol. I, chap. IV; p. 150 n.2; cf. Aristotle, *Politics,* 1256b 30–1257b 40.)

Marx has his reasons for wanting us to see our own image in this ancient anticipatory portrait, as if Aristotle had sensed the oncoming dangers of a possible society become frenetic with the search for unlimited, and hence meaningless, chrematistic wealth, rather than the intelligent search for the human good.

We have stepped, with Marx, a little distance outside the development of his formal theory in order to gain perspective on its implications for human life. If capitalism is indeed a set of social relations in which we are ourselves immersed, we are likely to need all the help we can get in comprehending our own situation. To this end, Marx relates a short, ironic narrative.

A free bargain has been struck, he says, like any other bargain in the marketplace, between a worker and his prospective employer. The worker at this point is "free" in two senses, which Marx distinguishes: he is "free" to strike whatever bargain he chooses, but he is also "free" of a crucial possession—he is "free" of any ownership of the means by which he might work. Within the terms of this double-edged "freedom," he strikes a fair bargain: he will be paid what it costs to feed, clothe, house, and otherwise provide for himself and his family; this is the cost of reproduction of his labor power. In return, his new-found employer will own his labor power. On the conclusion of this transaction, we leave the free marketplace:

Accompanied by Mr. Moneybags and by the possessor of labour power, we therefore take leave for a time of this noisy sphere where everything takes place on the surface and in view of all men, and follow them both into the hidden abode of production, on whose threshold there stares us in the face: "No admittance except on business." Here we shall ... at last force the secret of profit making.

This sphere that we are deserting, within whose boundaries the sale and purchase of labour power goes on, is in fact a very Eden of the innate rights of man. There alone rule freedom, equality, property and Bentham....

On leaving this sphere of simple circulation or of exchange of commodities, which furnishes the "Free-trader *Vulgaris*" with his view and ideas, ... we think we can perceive a change in the physiognomy of our *dramatis personæ*. He, who before was the money owner, now strides in front as capitalist; the possessor of labour power follows as his labourer. The one with an air of importance, smirking, intent on business; the other, timid and holding back, like one who is bringing his own hide to market and has nothing to expect but—a hiding. (Vol. I, chap. VI; p. 172)

Overdrawn, perhaps, angry and caustic as this little caricature may be, it forcefully epitomizes the class relation that emerges from the economic analysis: on the face of it, a fair bargain in the free market so much celebrated in the mythology of capitalism, but beyond the marketplace, the fact that possession of the means of production is divided in such a way that the labor power of the one must inevitably become the possession of the other, and hence, no actual symmetry in the "free" bargaining process.

This is not only a question of "impoverishment" or of a bargain which is unfair in the system's own terms: it is rather a problem of the status of human work. The worker whose product is another's is inherently alienated from his own product;

and if he is alienated from that product, which by social and legal right inherently belongs, as a consequence of the bargain, to another, then he is *a fortiori* alienated from the work itself. This is the alienation of human activity.[10]

We, as members of corporate society, schooled from our earliest years in its ethic, take this altogether for granted—we expect to be free *outside* the workplace and have no concern for freedom of our work itself providing it is pleasant, "rewarding" (as we say), and leaves us at large during normal spare time and vacations. We have perhaps no conception of "free" work in Marx's sense. We do not perceive the class structure that causes us to make the unequal wage bargain, and we are rather affronted than otherwise to have it pointed out to us. We feel we are paid fairly, perhaps generously, for our working hours; Marx is simply showing, formally and carefully, a sense in which we are not. We are paid for the hour equivalent of the car, the home, the food, the clothing, the amusements. That is a certain fraction, calculable though not normally accounted for in our society, of the working time we spend on the job: a fraction which Marx calls the "rate of surplus value."

There are many ways to think of this rate of surplus value, and many ways in which it can be adjusted overtly or covertly in the struggle—implicit or explicit, but under capitalism inherently never-ending—over the wage contract. Marx examines these transformations with great care; here it is sufficient to point out that the rate he has in mind can be thought of as a

---

[10] The term "alienation" (or "estrangement," German *Entfremdung*) is very important in the early writings of Marx, in such essays as that on "Alienated Labor." It is a classic question among scholars whether Marx had fundamentally changed his mind by the time he wrote *Capital.* I think that, while he may have avoided the term to adjust his vocabulary to new tasks and a new readership, the concept remains fundamental to a reading of *Capital,* and there is no abandonment of the earlier position in the later work. Rather, it is applied in new ways to a specific task in hand.

division of the working day: so many hours for the reproduction of the labor power, so many hours of surplus value to the employer. But it is equally true that every minute of working time is divided in that same proportion.

In studying capital, it is essentially, according to Marx, a structure of social relations that we have to examine. Marx has to keep reminding himself as well as the reader, one feels, that it is not persons (despite his occasional caricatures) but types, in the sense of objective social processes, that we are exposing: they take individual form, certainly, in many ways, but the real object of analysis is a social structure of which the capitalist is, Marx says, simply a "representative."

At this point, I want to make a suggestion whose roots I think are hinted in Marx's text: the "capitalist" Marx is describing is often not a person at all. Marx sees very far into the future of the careening system he is describing, but we have come a hundred years further down that road, and the logic has carried itself to a phase he foresaw but could not himself fully contemplate. Today we can see that the "capitalist" Marx describes has become above all the corporation, that artificial person given legal license during the nineteenth century to enjoy immortal life, and to aggregate resources and move with a will vastly beyond that of the boldest member of any aristocracy of personal wealth.[11]

---

[11] It is hard to appreciate adequately the significance of the legal processes that brought the corporation into our midst, as a dubious quasi-citizen of the polity. It was the "Dartmouth College Case" (1819) in which the Supreme Court took a decisive move in this direction; Chief Justice John Marshall wrote these awesome words: "A corporation is an artificial being, invisible, intangible, and existing only in contemplation of law. Being the mere creation of law, it possesses only those properties which the charter of its creation confers upon it.... Among the most important are immortality, and, if the expression be allowed, individuality...." (*The People Shall Judge*, Vol. 1, p. 460).

The *human* capitalist Marx describes is torn: he enjoys pleasures, with the result that use-value vies with exchange value,
consumption vies with accumulation. As capitalist, he accumulates with a rigor inherited from Calvin, but he is only human
and, like Faust, Marx says, he finds that "two souls ... dwell
within his breast"—a certain "fellow-feeling for ... Adam" corrupts the logic of accumulation with which he is entrusted
(Vol. I, chap. XXIV, sect. 3; p. 556).

It is not so with the *corporation*, which as an impersonal
being is able to carry through the logic Marx describes to a
pure result: this deathless, artificial entity—this Leviathan,
which in Hobbes's time appeared to be the State—makes, *can*
make, no compromise with the logic of accumulation. It is,
then, with the corporation that most of us enter into our contract of employment.[12]

Marx is quite blunt about what this means, which is that
those of us who accept employment have sold ourselves. We
are no doubt very happy to have done so, since the alternative, un-employment, is fraught with suffering and despair. We
freely make our minds and our skills into commodities. We
despise ownership of another person, which we rule out as

---

[12] It is worth turning to Hobbes's account of Leviathan, which one
can't help but feel Marshall must have been reading: "This is the
generation of that great LEVIATHAN, or rather, to speak more reverently, of that mortal god ... *one person, of whose acts a great multitude,
by mutual covenants ... have made themselves every one the author....*"
(*Leviathan,* chap. 17.) Hobbes of course is characterizing the *state* as
he understands it, but the concept has now far outrun his intentions.
Marx, already in his time, is deeply impressed by the corporation
(the "joint-stock company"), which he sees as a necessary stage in the
transition to ownership by the producers themselves, i.e., as directly
social property. He speaks of it as "an abolition of capitalist private
industry on the basis of the capitalist system itself." (*Capital,* Vol. III,
p. 570.) A similar perception of the corporation as the socialization
of the institution of private property was expressed by Adolf Berle in
Berle and Means, *The Modern Corporation and Private Property.*

"slavery," but without complaint we sell our working lives by segments to others we call "employers." In this sense we make ourselves slaves for the best part of our lives, the working core.

Here the wages or the salary, the nature and conditions of the work, are not the central point; it is rather the *status* of our human activity—as our own, or the property of another—of which Marx speaks primarily. If we are going to be slaves, it is nice, surely, to be well-cared-for house slaves with interesting tasks and many hours off around the fringes; but slavery is slavery. The central question is of the structure of the working relation, which under capitalism is, in principle, ownership of one's labor power by another and hence alienation of one's human activity. Better pay and all the amenities of attractive conditions of work become, Marx says, "golden chains" that bind us all the more firmly in servitude. It will not fit into our vocabulary to envision that "work" might, in a different system, become a domain of freedom.[13]

To return to the development of Marx's formal theory, we now have seen that the way from M to M' is through the employment agreement and the consequent claim on surplus value. It is now necessary to look more closely at that process. We have written:

$$M—C—C'—M',$$

---

[13] Throughout the essay on "Alienated Labor" runs the principle that our work ought to be free, and that our free activity is our real work, "...for what is life but activity?" Our "free, conscious activity" is our end in life, which is denied under alienation. The term "spontaneity" is often used by Marx in the earlier writings to characterize the radical freedom he is talking about: our activity in our work *ought* to be our "own spontaneous activity." Marx is, again, very close in spirit to Aristotle in seeing free work as an end in itself (as in *activity*, *"energeia"* in Aristotle), whereas alienated work is a mere means, and external. Admittedly, with the introduction of the *free will*, the concept of "freedom" may have undergone a sea change between Aristotle's time and ours.

where C initially denotes that magic commodity, labor power, which the capitalist purchases in order to generate surplus value. More generally, however, C must denote all the elements he must purchase in order to initiate a new cycle of production. Labor power must in any case always be among them, but the capitalist must also acquire the means of production to put that labor to work. In fact, by definition, the "capitalist" alone, through past accumulation, possesses money sufficient to purchase the means of production that will serve to put labor to work. So C really has two very different components—labor power itself, and the means of production.

Each of these categories of the capitalist's investment consists of commodities whose value is measured by the labor that has gone into their production, that is, labor power, whose value is the cost of its reproduction, and raw materials, factory buildings, and machinery, which are the products of earlier cycles of production. But these two categories function very differently, from the capitalist's point of view. The first is an investment in *living* labor, which will create new, expanded value; the other is a purchase of *congealed* labor, which can do no more than pass its own value on, proportionately, into the new product. These Marx calls, respectively, *variable* and *constant* capital. The latter he calls "constant" in the sense that, being already complete, it does not, like "variable" capital, grow. Constant capital then consists of "dead" labor which the entrepreneur brings to living labor in order to make new production possible:

> Capital is dead labour that, vampire-like, only lives by sucking living labour, and lives the more, the more labour it sucks. (Vol. I, chap. X, sect. 1; p. 224)

One need not, perhaps, phrase it quite so vividly, but the principle is striking. We might say that M drives forward to M' as rapidly and efficiently as possible; it achieves this goal, which for capital can be its only goal, in setting current labor to work on the means of production, which are in turn the products of past labor. The means of production, apart from living labor, can produce no value.

What then does capitalism "produce"? In the defining formula, $M—M'$, we have seen that the product in the ordinary sense cancels out. By means of products, capital in the strict sense produces new accumulation: that is, the system reproduces itself, or, as we should say, the product of capitalism is the system itself, reproduced on an ever-increasing scale. Capitalism thus tends to expand into any sphere that had previously escaped its grasp, such as the family farm or the individually owned store or business, and it inherently seeks to press its way into any new areas of the world which afford an "opportunity" for investment. Whether it is a question of a retail chain or the third world, capital's concern can only be that passage from M to M'. The personal capitalist may have compromised with human concerns, but the ultimate capitalist, the corporation, bearing legal responsibility to its own stockholders, must ultimately justify every significant act in terms of prudent regard to return on initial investment. $M—M'$ is a principle built into today's corporate law, as every alert director keeps constantly in mind, or overlooks at his peril.

Circulation and the Market [14]

The story of capitalism has to be told on two levels. The first is that of the individual entrepreneur, his production process, and his contract with labor; the second is that of society as a

---

[14] In general, the question of circulation is the topic of Volume II, where Marx draws upon the work of François Quesnay, who developed "tables economiques." Marx's tables balance the overall exchange between sectors of the economy and thus become most interesting models for the kind of planning of an economy which he visualizes for the rational society of the future. (*Capital,* Vol. II, pp. 468–97, on "simple reproduction"; and ibid., pp. 565–99, on "reproduction on an expanded scale.")

whole, the social relations and processes that arise from myriad individual transactions, and the final economic balance. On the individual level we meet the battle of competition and the sometimes fierce, always-thrusting individual drive that is the genius of capitalism and the motive force from which all other motions follow. Yet the perspective at this level is limited; it gives rise, in ways Marx describes, to the system of illusion under which "profit" seems merely an obvious return on "capital." From the individual standpoint, capital cannot be seen as the result of an intense and very recent social development. The laws of motion, by which those who think of themselves as agents are in fact caught up in cycles of greater social motions that govern them like forces of nature itself, remain invisible.

Marx has begun at this level, examining individual transactions and the individual production process, and there he has found, he is confident, the secret of capitalism in the expansive principle of surplus value. Gradually, however, he moves toward ever-larger views of the marketplace and the interaction of capitals in the overall circulation cycle, which will become the explicit subject matter of Volume II. In this transition, he looks insistently for the relation between the individual drive and the social outcome, between the domain of the separate and the domain of the social, between the region of detail and vagary and the larger region of regularity and social perspective.

Each capitalist produces with the goal of sale in a specific market. There, his product enters into an exchange process in which, as we have seen, a continuous scrutiny is occurring, with the market appraising, perhaps on a daily or hourly basis, the socially necessary labor time incorporated in each item. From that arises, as a social judgment, a market price.[15]

---

[15] As we shall see, the outcome of this market evaluation is not that the sale price is literally equal to the "value"; things prove much more complicated than this, but as a component of the sale price, the judgment of *value* by the market is nonetheless presupposed (p. 252).

Since this price is a social average, if the individual capitalist can find any way at all to produce the same product
more cheaply in his own particular factory—that is, if he
can reduce the labor time incorporated in each item below
the social average—he will be able to realize a higher-than-
average rate of surplus value. He will thus gain a significant
differential advantage over the market. Typically, he may do
this by introducing some new machine, or by refinement of
production that increases the productivity of his labor. Hence
arises the compelling drive to analyze and improve the production process, or to expedite distribution and marketing
in any way possible so as to reduce turnover time. This is the
relentless force that was already dramatically transforming the
factory and the means of transportation and communication
in Marx's time—and is, perhaps even more dramatically, still
doing so today. Marx stands in genuine awe of the social gains
this energy has made possible.

Not only does competition drive each capitalist to achieve
whatever differential advantage he can contrive, but it insures
as well that every other capitalist will follow suit as rapidly as
possible. Each new invention, each advance in the production or circulation process, must spread like a wave through
the industry in a common effort to share in this new gain—
and of course, as this occurs, the ever-watchful market will be
reckoning the new socially-average labor time incorporated in
the product at its new reduced level. A new market price will
result, all differential advantage will be lost, but in the end the
product will be cheaper and society will have registered a gain.
Immediately, however, each capitalist will return to analyzing
his process for opportunities to gain a new differential advantage and restart the cycle.

From this social perspective, one of Marx's principal (and
most contested) insights now takes shape: the net effect of such
cycles of improvement in the productivity of labor is to drive
living labor out of the production process wherever possible,

and to replace it with congealed labor in the form of machines and other forms of constant capital. From his individual standpoint, the capitalist cannot see that living labor is in fact the only source of surplus value. Thus, while each capitalist makes important transient gains during the interval in which his differential advantage lasts, the net effect for all is a lowering of the only real source of gain, which is living labor. In the limiting dream of total automation, there would be no source of surplus value, the social structure would cease to reproduce itself, and the system of capitalism would dissolve. This vision demands more careful formulation and criticism, of course, but it may serve for the moment as an example of a certain process. Not only an overall social result, but a long-term, historic law might arise from the unseeing, immediate individual interactions of marketplace competition.

As a city looks different from the air, the detail of daily life receding and relationships emerging which could never be apparent from the ground, so Marx's synoptic view of the production and circulation processes as a single social whole gives us access to new concepts and conclusions. Only from this higher ground can we see the market itself as simply an incident in the production and reproduction of social relations, indeed of society itself, which in this cycling of its greater year is taking new forms, shaping within itself new powers, contradicting its past, and giving birth, perhaps, to some very different future. The individual point of view sees the entrepreneur, as owner of resources, encountering the worker, who having no such resources is necessarily seeking a job. In that view, their relationship is taken as given: it appears in this sense as merely "accidental."

From the social perspective, by contrast, we can see that the case is very different. The same relationship appears as an outcome of previous cycles of accumulation, as itself a product of the system. It is not even necessary to go back to a distant past, to trace the history of the great capitalist fortunes, or to envision some original act of labor or of initiative in a "state of nature," as the origin of the fact of property.[16]

Rather, in the social perspective, we can see the birth of the property relation reenacted before our eyes. Whatever an original fortune may have been, that first increment is soon a merely evanescent quantity with respect to the regeneration of surplus value and its accumulation which is occurring on a continuous basis. As new accumulation eclipses the old, what we see is the effect of the mere claim one segment of society has on the labor of another, by virtue of the institution of property in the means of production. We witness a perpetual separation of the social product into two portions, one going to the reproduction of the labor force, and the other to those who hold social claim to ownership of the means of production. "Capital" thus lies before us, not as a collection of *things,* or even as a static social relation, but as an ongoing *motion,* which is continuously reproducing the social relationships of a hierarchical society.

That sense of ongoing motion is perhaps nowhere more vivid in *Capital* than in the monumental Chapter 15 of Volume I, On "Machinery and Modern Industry" (pp. 351ff.). A reader who undertook no more than this chapter might get some sense of the magnificent shaping energies of capitalism, and the ambiguity of their outcome: new, unheard-of productive forces emerging on an almost daily basis, while at the same time what seems on its face a degradation of the role of man the worker, who was so recently still a craftsman and owner of his own tools, and to that extent, at least, still autonomous in the production process, comes about.

---

[16] Locke envisions, in a kind of myth of the state of nature, the origin of the right of private property in an act of individual labor (*Second Essay Concerning Civil Government,* Chap. V). Rousseau similarly refers the question to an imagined original state but sees this primal claim to a right of private property as an outrage against mankind (*Discourse on the Origin of Inequality,* Part I). These views are part of the discussion of WEALTH in the *Syntopicon* (*GBWW,* Vol. 3, pp. 1038–49).

Marx describes in this chapter a vast concentration of resources, inexorable refinement of the labor process, and the substitution, at every turn, of machinery as surrogate for the human hand and human skill. Scale is decisive: large capital drives out small, centralization replaces scattered enterprises, transportation is revolutionized. Credit systems transform the market process into instant, ideal transactions that pare turnover time to the minimum—and make possible a concentration of investment in "cyclopean" machines which man can do no more than tend. Marx traces with fascination the development of machines that make machines, since even the human machine-maker appears as an annoying bottleneck in the burgeoning development of the industrial age.

All this can be said in a spirit of wonder and praise, since for Marx it is a vision of the almost unbelievable powers of man when those powers are organized on a social scale. Everyone wonders, of course, at the machines, and the rationalization of the production process; but from Marx's perspective such a rationalization holds meaning which others seldom articulate. On the individual scale we saw competition, the restless forces of private gain. On the social scale we see, by contrast, coming out of this, aggregations of capital, and *de facto* social organization and cooperation, however unwitting and unintended, on an unprecedented scale: the power of social labor. It is science and technology, of course, that in one sense make all of this possible. But science and technology are inherently social functions themselves, in which each investigator in due course shares in the gains of all, and can only work through access to the unfolding social product.

The interweaving of products and processes, circulating on a new, world scale, represents the construction of a totally new form of society. Competition and the accidents of the marketplace retreat in importance before the fact of unprecedented social organization and *de facto* rational cooperation. Capitalism reveals itself as giving birth in very fact to its opposite: competition is continuously generating unrecognized cooperation. Yet this new society-in-process, while it remains in the bonds of a hierarchical social order, is seen by Marx as fraught with deep contradiction: machines that save labor only cause us to work longer and under greater stress; machines that are works of great intelligence drain work itself of any intelligent involvement. Later, we will consider the implications of this double oracle: prospect, and denial.

Marx dismisses the usual analyses of the economic process given in terms of "supply and demand." This is not, of course, because of any lack of appreciation of the importance of the free market as the foundation of the system of capitalism: in all his analyses, Marx presupposes the free market. It is the free market, indeed, that continuously determines the values that underlie exchange. This is as true, we shall see, of the markets in capital itself, in money, and in land, as it is of those in labor power and the products of labor. But "market economics," though it has its use in relation to the immediate phenomena of the marketplace, does not yield that vision of the overall movements, laws, and problems which makes of economics the science Marx is seeking. Only the larger view, from which the transient phenomena of the market cancel out, can do that. It is the larger laws, not the passing turns of the market, which are Marx's concern in *Capital*.

Similarly, in place of the bourgeois term, "demand," Marx speaks of "social need." He defines this as the effective purchasing power of society as a whole—total effective demand—as backed up with the money necessary to effect a purchase. This is a somewhat different concept from the "demand" often thought of in connection with market economics, whose point of view is first of all that of the individual market transaction. Individual workers may need or "demand" all sorts of things, certainly far more than society is prepared to offer; but even when these desires are supported by occasional access to cash, and hence become "effective demands" on an individual basis, they must remain irrelevant to the overall economic accounting, except as vagaries which over time are canceled out.

Social need is a class concept, and on balance only such a class approach to the market has significance for the formal science.

The social product is allocated to the working class, who are the bulk of the consumers, on the basis of the rate of surplus value; the drive to increase profit insures that this rate will be kept as high as conditions permit, and essentially a stable quantity. Therefore the flow of value into the category of purchases by workers, the reproduction of variable capital, is quite strictly determined. It consists of the basket of provisions required to reproduce the worker and his family under the conditions, including accepted standards of social expectation, in any given social era. *This* is "social need," and we can see why it is the operative concept in any really synoptic view. Its detail, in terms of brands, styles, fads, and the like, is indeed a matter for market determination and a question of prime concern to producers, investors, and advertisers—the immediate participants in the competition for sales. Over longer periods, because of the overall cheapening of products with improved methods of production, its content tends to improve. But its total amount in any given era is limited by the requirements of capital to command surplus value, and its basic composition (housing, clothing, nutrition, transportation, education, amusements, etc.), since it must meet fundamental human needs, can vary in any one period within only rather narrow limits.

Profit on Capital[17]

We now move into Volume III of *Capital*, and at the same time we descend to a dramatically lower level of abstraction. We move, that is, closer to the detail of the world of capitalism, a world that, as Marx argues in this third volume, systematically hides its own essence in illusion and mystification. We began in Volume I by identifying the abstract elements of capitalism,

---

[17] *Capital*, Vol. III, pp. 117–99, 241–306.

in the commodity, in labor value. and in the secret of surplus value. As we move away from these, we will be reconstructing the phenomena of the daily life of capitalism, tracing always as we go their relation to that veiled essence that alone can render the phenomena intelligible.

It is ironic that *Capital* as generally known—as reprinted, for example, in the *Great Books of the Western World*—consists only of the first volume of the three Marx wrote. The result is that most readers are left at that initial level of abstraction, which cannot speak the full truth of the world in the world's own terms: it tells of the secret and essence of capitalism, but it has not yet taken on the crucial burden of working back to the ordinary events and familiar discourse of our economic lives. That is the special task of Volume III. The importance of this is not only to make the connection in such a way that the theory of Volume I can actually be interpreted and applied, but also to illuminate the origin and nature of illusion itself. Marx traces here the systematic generation of false consciousness and the process that has since come to be called "reification," by which entities take shape in consciousness which have no objective counterparts in reality.[18]

This must be a dark study, not only because it descends ever more deeply into shadows of distortion, but because a certain note of despair inevitably intrudes as it becomes apparent how total and almost indefeasible such systematic social illusion may be.

The first step in this descent is the transformation of the *rate of surplus value* into the *rate of profit*. This distinction arises

---

[18] The term "reification" is developed by György Lukács, one of the most interesting interpreters of *Capital,* in his *History and Class Consciousness.* See, for example, the essay "Reification and Class Consciousness, I: The Phenomenon of Reification," pp. 83–110. Marx uses the term "fetishism" to characterize this conversion of a relationship into a thing; this is already apparent in the discussion of the commodity, but Marx reflects on the process in the third volume (*Capital,* Vol. III, p. 516).

because capital committed to an enterprise and tied up in it, as machinery, raw materials, or buildings, demands recognition. Capital, as such, contributes nothing new to the value of a product. But tied up as constant capital, it is prevented from utilization in other forms, in which it might be directly employing living labor and hence reaping surplus value. As capital, it is "owned," and no "owner" will forego the opportunity to receive surplus value unless he receives adequate compensation. In fact, since his capital is crucially needed—for a certain minimum of constant capital is always required to put living labor to work—it might seem that he could demand whatever price he chose; but in fact, there is a limit.

Since the only source of any new value is the surplus value created in the labor process itself, the owner of constant capital can demand for it no more than a share of that new increment of wealth. The measure of this "share" is in turn determined by exchanges in a new kind of market that arises, *a market in capital itself*. In it, capital flows to investments that offer greatest return, with the result that overall, and to the extent that this market is universal and fully effective, all capital, constant or variable, tends to receive the same rate of return. "Profit" becomes the ratio of surplus value to the *total capital* invested in an enterprise. This will be less than the rate of surplus value, in which the numerator was the same, but the denominator was variable capital alone. The rate of profit will thus be lower than the rate of surplus value, as total capital is greater than variable capital, which directly employs labor.

We can see that the market in capital, which itself represents a major historical development in the unfolding logic of capitalism, must greatly alter the distribution of returns to investors from the production process. In the case of an industry that utilizes relatively small amounts of constant capital, the rate of profit would approach that of surplus value itself, its upper bound. On the other hand, a process that requires large amounts of constant capital will generate relatively little surplus value in relation to the capital invested—it will have a low rate of profit. However, through the operation of the capital market, neither investor will actually see the rate of profit his own enterprise would have generated if it operated alone. All industries are closely coupled through the capital market, and capital will thus flow out of the more profitable enterprise and into the less, until finally under the pressures of the competition of capitals, money invested in each becomes equally rewarded. As it thus comes to share in the average rate of profit, it continuously transfers surplus value from the intrinsically higher profit enterprise into the lower.

The effect of the capital market is thus to socialize the exploitation of labor: transfers of value occur constantly, through that market, from high-return to low-return industries, until all share in a common profit rate. This social process is not seen as such by the participants, but from the perspective of Marx's theory, we can now see how it would follow that in this way each capitalist would indeed become, as Marx puts it, "a shareholder in the whole social enterprise"—all are participants in a general "freemasonry of capital." [19]

Since the operation of surplus value is, from the outset, unseen by the capitalist, he is even less aware of its transfer, and he does not perceive the immense and crucial social process in which he is taking part. To him, it seems that each capital investment itself generates its own expansion of value. What is in fact a social result, through which the highly technical operation of the market in capital has become generalized and effective on a world scale, occurs altogether "behind his back."

Since, with other things equal, capital tied up in an enterprise is rewarded in proportion as the turnover time of the process can be shortened (thereby absorbing more living labor in any given period of time), the production process is constantly being driven at the highest possible velocity. This calls for shortening of the circulation time whenever possible, the invention of methods of credit which circumvent delays

---

[19] *Capital*, Vol. III, p. 312.

in marketing, and pressures on commercial capital involved in commodity exchange. Communications, illustrated in the nineteenth century by such advances as the introduction, first, of packet ships, and later, of the telegraph and the ocean cables, are driven by the competition of the capital market: the effort to gain differential advantage through earlier access to market information, or to realize a higher-than-average rate of profit through an edge in turnover time. It is only the average rate of profit in an industry as a whole, over time, that the capital market establishes. Hence despite the "freemasonry of capital" that is thus ultimately assured for the industry as a whole, the individual entrepreneur still feels the same pressure as ever to gain any temporary advantage he can contrive.

The illusion that capital—which in fact simply *demands,* by virtue of its status as private property, to be rewarded—creates the value which flows to it, yields corresponding distortions throughout capitalist thought and accounting. Bookkeeping, whose art is the new rhetoric of capitalism and hence of our era, is conformed to profit, not to the underlying surplus value. Everywhere that capital, in whatever form, enters the production process, capitalist bookkeeping imputes to it the general rate of return on which it insists. The accountant makes this entry with no sense of duplicity or distortion, but rather in due recognition of what he has learned to call the "cost of capital." In his view, the "cost of capital" is no different from the "cost of labor" or the "cost of fuel": each he sees as a component of the "production price" of the product. Ultimately, this becomes incorporated in the price of the commodity, and the consumer pays for it accordingly, again with no more qualms than were felt by the accountant.

We see now that the actual commodity markets of capitalism are not, as we had assumed in Volume I, assessing the *value* of the products in which they trade, but rather a *cost* based on a proportionate imputation of value to all elements of capital that have been involved in their production. Cost in this sense displaces value in all capitalist markets, and in economic theory as well. Price tends not toward true labor value, but toward a construct of capitalist bookkeeping that has the effect of transferring value in streams from high-profit to low-profit enterprises. We see now that the entire market operation, and with it, all thinking about the economic process, is systematically mystified in such a way that value and the source of value become invisible, just as the flows of value and the socialization of that flow remain unobserved.

Does this mean that, in practice, Marx's analysis, however true it might be at some level of abstract principle, has no real applicability? The question would seem to call for a thoughtful pause. We can see one thing, however: no matter what the participants themselves may think, if Marx was right in the fundamental analysis of Volume I, then the law of value still governs the overall exchange. There is no source of profit other than surplus value, and the analysis of Volume I remains decisive: total profit in the entire market, having no other source, must finally equal total surplus value. Hence, other things equal, it is as true as before that the rate of surplus value is the underlying index that governs the return on all capital: it is only the purchase and exploitation of labor power that originally generates the returns in which, by whatever route, all ultimately share. And if that is true, then it is at least as important as before to recognize the fact, and to be able to penetrate the mists of illusion, to see through to the engine at the heart of the machine. One prime mover turns all the wheels, however they may imagine that they turn themselves. All that was said at the higher level of abstraction in Volume I remains true now, but its elements operate in Volume III increasingly behind the scenes, and the burden of tracing their operations through the hieroglyphs of economic practice becomes the more onerous.

What becomes apparent is the emergence of a systematic illusion, and we see at the same time how even that illusion itself is a necessary product of the very system that it serves to veil. We are speaking, not of a merely technical error, but of a coherent myth, a fundamental inversion of perception of what

is going on in the universe of out experience. Capitalism constitutes an ever-ramifying *social* fabric, while at the same time
it presents itself as individualistic and planless, indeed the very
product of unfettered human nature.

Interest on Money[20]

We now turn to the consideration of *interest,* and with it,
take one more step downward from the clarities of the original
abstract heights into the confusions of the realm most familiar
to us, the daily phenomena of capitalism, with all its vicissitudes. We all know, as if it were one of the facts of nature, that
money "draws" interest, and we view with concern anyone who
keeps money in his possession for any length of time, without
putting it "to work." With interest, Marx says, capital takes its
ultimate fetish-form, in which it appears as if money itself were
capable of generating value.

Interest is, of course, a pivotal phenomenon of capitalism, the outcome of its relentless effort to achieve the power
to aggregate resources, through channels of credit, on short
notice and into ever-larger concentrations. Yet, by contrast
with the rate of profit, the rate of interest obeys no inner law.
Interest and profit must both, indeed, be carved from the
same source in surplus value; a division must in some way take
place between them. But for the determination of this point
of division, there is here no rational principle. Competition,
tradition, and the mood of the marketplace shift interest rates
in a never-ending contest between lenders and entrepreneurs.
The illusion that arises with this division of function between
profit and interest runs very deep. It seems as if the allocation
of surplus value were measuring something fundamental in
the economic process, as though two different *sources* of value
were being tapped. and rewarded on some proportionate

---

[20] *Capital,* Vol. III, pp. 459–573.

principle. "Interest" appears as a return on ownership *per se*, as though money tied up inherently generated a return. With this development of the capitalist market, it seems as if the original dream had been realized: money in itself seems to bear fruit, as a pear tree bears pears. It is as if interest "grew" from money. Marx calls attention to the Greek word for "interest," *tokos,* which first means simply, "offspring." [21]

It seems that the underlying formula of capitalism, $M—M'$, has now been realized without the annoying intervention of any production process whatever—that is, without the distraction of having to produce and distribute any product, in accord with the original more cumbersome, expanded formula,

$$M—C—P—C'—M'.$$

In turn, with interest now conceived as return to ownership *per se*, the function of the remaining profit of enterprise becomes seen as the return to a distinct function, as "wages of superintendence"—this despite the fact, which Marx and Adam Smith alike point out, that these functions of management are regularly in fact delegated to supervisory personnel, who as such have no share in ownership. Marx goes on to point out that the success of cooperative factories demonstrates that the capitalist's supposed role in actual production can be eliminated without its absence being especially noticed. [22]

Capitalist bookkeeping, which as we have already seen gives canonical form to a developing structure of illusion, now makes distinct entries: *interest* on borrowed funds, and *profit* on invested capital. The former appears as the "cost of money," and may be duly registered as a separate item even if the capitalist has advanced the funds himself, and no money has been borrowed. Confirmed in this perception by a coherent system of bookkeeping with its own terms and strict rules,

---

[21] *Capital,* Vol. III, p. 517.

[22] *Capital,* Vol. III, p. 517.

the capitalist can hardly be held responsible for his error in imagining that money in itself contributes value to his product. Money is rewarded with interest in recognition of this magic efficacy to contribute new value by virtue of the mere passage of time—"asleep or awake, at home or abroad."[23]

Preserving its myths, this world of illusion is intent on what becomes in effect a deliberate effort at self-deception: reproduction of its own ignorance of the realities on which it in fact rests.

Rent [24]

We now meet a yet further exaction from the production process, in the form of rent. Rent, Marx says, arises because "a monopoly to a piece of the earth enables the landowner to exact a tribute" in exchange for permission to use the land for the production of crops. There is a difference, however, between the "tribute" demanded by the landowner and the tolls taken by the capitalist and the banker. Although a demand is made for a return that does not arise from any actual contribution to the value of the product, at least in these two cases there is some basis in actual value: capital and money both represent value, definite amounts of crystallized labor time. When the landowner demands a payment for the use of unimproved land, however, his claim is not based on any actual value whatever: since it is not a product of labor, land *per se* has no value. Interest and profit represent returns on actual values; rent on unimproved land is a return on no value at all.

---

[23] *Capital*, Vol. III, p. 517.

[24] *Capital*, Vol. III, pp. 751–87; 882–916. Note that these readings do not address the question of "differential rent," which in the overall theory becomes very important as a development of the "marginal" principle, but which it has not been possible to include in the present essay.

Any "value" that is set on unimproved land is, then, a purely irrational expression. Increasingly, as we move with Marx from the original surplus value through capital and interest to rent, we see social relationships of ownership distorting and veiling the actual underlying economic process, twisting the economy through the rhetoric of bookkeeping into false measures of cost and price founded on arbitrary demands, detached farther and farther from the only source of value itself, human labor.

Marx is looking in particular at the English landowning class as he writes about rent (specifically, "ground rent"), but he points out as well that the same account could be given in other areas of an economy, such as mining. Because he founds his analysis systematically on first principles, it is not bound to the institutions of his own time; "rents" on unimproved properties in various forms are to be found throughout our contemporary society, and everywhere they occur, they are subject to Marx's finding of irrationality. Landed property, Marx says, "presupposes that certain persons enjoy the monopoly of disposing of particular portions of the globe as exclusive spheres of their private wills."[25]

Agricultural land is certainly a prime instance of this, but any "portion of the globe" that has a use-value and can be "chained down" is a candidate for a rent—namely, a demand for payment for its use.

If we ask how large this rent might be, we see right away that we are confronted with an extreme case of the situation we met before. Once again, the only source of surplus value is the labor process; again, the question is that of division and distribution of the surplus value achieved there. The capital market decided the allocation to capital in the form of a rate of profit; the demand for interest yielded a fundamentally arbitrary division between the banker and the capitalist. In each case, however, the return took the form of a *rate* calculable in

---

[25] *Capital*, Vol. III, p. 772.

relation to an actual value as denominator—the labor time that is crystallized in the capital or the loaned funds.

In the case of rent, there is no such denominator. Rent, as a return on what has in principle no value, is not a rate but a social exaction, worked out as an implicit or explicit confrontation between landlord and capitalist.[26]

Once again, bookkeeping plays its responsive role: where before it imputed a rate of return to all money as such, it now—where there is no value on which to base a rate—reasons in reverse and *imputes* a value. In effect, it simply conjures up the missing denominator. A value is computed for the land, which is thus "capitalized" to justify the rent as if the latter were a rate. This, according to Marx, is the origin of the supposed "value" of land—it is simply a rent capitalized according to an assumed rate of return. Because social relations have exacted a rent, a value is attributed to the property. The house of mirrors is now very nearly complete.

Once the fiction has been established that land has a "value" and a "price," then by a kind of second imputation, or inversion of the inversion, bookkeeping will regard the use of this now valued land as a "cost" and will impute "rent" to it as a return on its facade of value. This becomes a new component in all pricing and in turn skews all market judgments to include a component toll to landownership. Marx acknowledges that it is extremely difficult to keep one's bearings in this labyrinth of fictions, all of which are so worked into our common experience that they feel to us like the most evident of truths.

With the theory of rent, Marx has essentially completed the inquiry that began with the identification of that abstract cell form, the commodity. We have before us now a new panorama of the phenomena of capitalism, not as mere empirical patterns or observed "laws," but as evidences of deeper social processes. What looks on the surface like fragmented thrusts of individual competitors on the track of separate advantage, reveals itself as governed by deep underlying social relationships, and we

---

[26] *Capital*, Vol. III, pp. 772–78.

now see vividly before us social processes and cooperation just the opposite of the atomic market processes in terms of which capitalism pleases to think of itself.[27]

A new society has been forged and is being advanced and perfected on a daily basis. It is a society with immense social and technological powers to achieve the human good, were the old mythology to be shed, and its real possibilities recognized— understood and mastered to serve rather than to threaten and possibly overwhelm us.

What has been revealed most clearly and consistently, if Marx's analysis is correct, is that, despite every illusion to the contrary, all rewards in our system are generated out of one underlying process of exploitation. The term "exploitation" refers simply to the basic principle for Marx, that labor is never paid at the value it produces, while the difference, that surplus value, is systematically transferred to reproduce a hierarchical society. It is a system that runs, not to achieve the human good, but to reproduce an arbitrary structure of inequality. The formal analysis has shown us that, if its terms are correct, we live in a perverse social order which, despite its persuasive rhetoric to the contrary, and despite all its promise and all our efforts, runs inherently to produce and to reproduce a twisted world.

---

[27] The seed of this perception, that a larger, systematic social result arises from the operation of many individual actions, each of which considers only a separate advantage, can be found (like so many of Marx's ideas) in Adam Smith, who says of the individual entrepreneur, "…he intends only his own gain, and he is in this, as in many other cases, led by an invisible hand to promote an end which was no part of his intention…." (*Wealth of Nations*, Book IV, chap. 2.) But where Smith sees in this a *limitation* of our human abilities—we will make things very much worse if we try to plan for a social goal—Marx understands this as a dialectical *prospect*: we have powers we have not recognized or utilized. Marx speaks of Adam Smith's "invisible hand" in these terms: "…trade … rules the whole world through the relation of supply and demand—a relation which … hovers over the earth like the fate of the ancients, and with invisible hand allots fortune and misfortune to men…" (*German Ideology* [New York: International Publishers, 1970], pp. 54–55).

Marx has traced to their origins the illusions that he believes altogether prevent us from seeing this truth about ourselves. Is he correct in this belief? It is not a small question, nor one confined to economic or political issues. If he is right, we are wrong about the very nature of freedom itself, and of our humanity. We think we are free, he says, but we are not; we think we understand ourselves and what we call our "human nature," but if Marx is right, we are still only on the way to the comprehension of what it would mean to be fully human. We have not yet formed a rational and cooperative society, whose members know, in practice, effective freedom and actual equality of opportunity. Yet if Marx's analysis is at all valid, we might be much closer to this than we permit ourselves to recognize.

Transition to a Dialectical Theory: The "Trinity Formula"

Having traveled such a long road in the development of his theory of capitalism, Marx now offers us, as a kind of antidote to all the deceptive mirrors we have looked into, a magic mirror of his own in which the whole work, with all its contradictions, can be reviewed with a new sense of perspective. I have attempted to distinguish throughout my own account of Marx between the "formal" and the "dialectical" theories, the former being, in principle—though not in any literal or sequential way—"framed" within the latter. It has often been difficult to preserve that distinction, since, however they differ in principle and in their roles, the two run together throughout the work, the critical as a kind of subtext, we might say, to the formal. I think Marx wants to help us now in the transition from one mode to the other: from the "formal" to the "dialectical" reading of the work—and thus, in reflection, to put the formal theory in its dialectical frame. He does this in a remarkable chapter, in a sense the high point of Volume III or even of *Capital* itself, which he entitles, "The Trinity Formula."[28]

---

[28] *Capital*, Vol. III, pp. 953–70.

Here, near the end of the work, we are invited to reconsider the entire account with new insight.

The reference of the chapter's title is to the Christian trinity, which Marx likens now to the three great elements of the conventional theory of capitalism, the theory that corresponds to capitalism's perception of itself—what Marx regularly calls, the "bourgeois theory." He recalls these elements in the pattern:

$$CAPITAL / INTEREST$$

$$LAND / RENT \qquad\qquad LABOR / WAGES$$

where in each case an imputed "source" is paired with the mode of return that is assigned to it according to the distributive justice of this mythical order: thus, land "earns" rent, capital "earns" profit which generates interest, and labor "earns" wages. Marx says these elements are like the persons of the Holy Trinity: Father, Son, and Holy Ghost. Why does he say this? We know that he regards this bourgeois "trinity"—just as he regards the Holy Trinity—as a form of mystification. But why has he chosen this particular mode of parody? To arrive at an answer, we might remind ourselves that in earlier days, before *Capital*, when Marx was writing on these same subjects, he made clear that he saw the Christian mystery not simply as error. It is, he wrote, like capitalism itself, a mode of alienation—the Christian vision was not simply wrong, but an alienated perception of an important truth.[29]

---

[29] Marx discusses his understanding of the role of religion in early essays, for example, "On the Jewish Question" (*Karl Marx: Early Writings*, ed. T. B. Bottomore [New York: McGraw-Hill, 1964], pp. 3–40). Marx's famous characterization of religion as "the opiate of the people" does not mean that it is stupefying, but that it projects a vision of the human community in alienated—remote—form. Marx speaks in the same terms in *Capital:* "The religious world is but the reflex of the real world." (Vol. I, chap. I, sect. 4; p. 83.)

Christianity was seen then by Marx as an inverted perception of the vision of a longed-for human future—the Kingdom of Heaven as an alienated vision, projected into realms of eternity, of the liberated human community toward which mankind is striving in actual history. Behind the Holy Trinity, Marx claimed, lies a secret, and that secret is *man*. Where the religious consciousness has projected God, there lies in reality, hidden, a dawning recognition of the possibilities latent within the human community. The power of that veiled truth, refracted in one way and another, has been a primary moving force of Western history since the time of Christ.

In the same way, the capitalists' trinity, in Marx's view, contains a secret: behind the separate categories it asserts lies the one source of value from which they all flow—living human labor. This is so because, if Marx is right, "interest" and "rent" are, for all their separate pretensions, no more than shares in surplus value, which is what labor alone creates. This labor in turn Marx sees as having its ultimate fruition in human community—that which we are working for, he suggests, even if we don't realize it. Hence the analogy to the Holy Trinity is perhaps "not altogether fool," but a strong hint that behind the error lies an essential perception about ourselves and our society.

Of course the allusion to the Christian trinity in this connection is ironic, and boldly so. But irony is a mode that is inherent, and reveals itself in one way or another, in all dialectic. The path of dialectic lies through confrontation with contradiction. Marx collects the contradictions incorporated in the economic trinity, which we have seen along the way as the formal theory traced out the development of the three elements. There is in truth no relation between capital and interest, or between land and rent; and land itself has no value. How can such ratios be taken? Not by way of economic principles, which do not serve, but rather by way of social relations. The crucial error, underlying the apparent economic irrationality, is the social institution of private ownership of the means of production.

In the formula, "LAND/RENT," for example, the noun "land," which seems to represent a fixed and definite *thing*, is in fact a surrogate for a historical social relation, landed property, property in "a portion of the globe." Similarly, "capital," which sounds like a *thing*, in turn represents a historical social relation which establishes certain persons as "owning" the means of production.

This substitution of a thing-word for what is in fact not a thing is the *reification* we spoke of earlier, and we might say now that the bourgeois trinity "reifies" land, labor, and capital, as necessary economic realities, where in fact by each of these ought to be meant a certain historical social relation— "ownership" in the case of land and capital; "nonownership" in the case of labor. The reified social relation, once recognized for what it is, links all the persons back to one center: living labor—not alienated "labor," the economic category to which "wages" are ascribed in the belief that it is thereby fully paid— but actual labor, which is never fully paid. This is the single source of all the categories.

We now see the relevance of the Holy Trinity. For the secret of that is the incorporation of the three "Persons" in one mystic unity. Correspondingly, behind the illusory distinction of the elements of the economic trinity there is an unseen principle of unity: the weave of economic society itself, which Marx sees as the social character of our labor. The weave of the markets in capital, money, labor, land, and all commodities, and the fine-tuning of densely intertwined production processes in every category, has yielded, as we have seen, immense advancement in these forces of social labor, in the cooperative processes of science and technology, communication and transportation, and techniques of social planning and organization on unprecedented scales. The nature of labor has been entirely transformed to match this *de facto* socialization of the production process; out of it are coming, Marx sees, the makings of a new community of mankind.

On these terms Marx has shown that his vision of the future is to an important degree already with us, and yet wrapped in a cloak of invisibility, in the misconceptions and alienation of the present. Marx is not given to laying out blueprints of the future. Readers often feel disappointment in this; they want to know his prescription for human society. He does not indulge in such speculation, not out of avoidance of a task that would be a heavy one to undertake, but on principle. If history advances dialectically, the future does not come to us in blueprint form. We make our ways into it, reading the signs and shaping our advances as we go.

The future appears to us in the very pages of *Capital,* as we come to understand the unfreedom and illusion under which we now live and begin to read the signs of a possible human future—already so nearly in our hands, and yet so far from our grasp. Under the spell of the "Trinity Formula," Marx permits himself some reflections on the future toward which it points:

> Freedom, in this sphere [of human needs], can consist only in this, that socialized man, the associated producers, govern the human interchange with nature in a rational way, bringing it under their collective control instead of being dominated by it as a blind power; accomplishing it with the least expenditure of energy and in conditions most worthy and appropriate for their human nature. [30]

This is not an idle wish, Marx would have us believe, but a reading of the evidence. Even in the bondage of the present "domination," if we read through the opaque symbols of our current order, Marx has shown us how many signs there are that we are already well on the way to being, in fact, members of a cooperative and rational society, with rapidly accumulating experience in "governing the human interchange with nature

---

[30] *Capital,* Vol. III, p. 959

in a rational way." Reality, Marx has endeavored to show us, is working in the direction of freedom; it is myth and illusion, and an anachronistic social order, that bar the way.

It would be the work of another essay to read *Capital* a second time, systematically, as a work of dialectical science, and to examine and develop the concepts which that phrase holds within it. It has been necessary here to walk once through the formal structure, noticing often enough as we went along the contradictions and illusions that suggest the further, dialectical reading. Here, in concluding the present study, we can only look briefly at this question, to consider what it would mean to speak of *Capital* as a "dialectical science," and to sketch what a dialectical reading might entail.

Three Stages in the Dialectic of the Dialectic[31]

We have spoken earlier of two models of "science"—that which aims to save the appearances, and was exemplified by Ptolemy, and another, exemplified by Newton and the *Principia*, which seemed the model that Marx follows in *Capital* in its aspect as formal science. Now, however, I am proposing a third, for I believe that the dialectical reading of *Capital* is not less "scientific" than the formal one, but rather that it very much enlarges the sense and scope of "science." We would be right, I believe, in speaking of *Capital* in its larger aspect as a work of dialectical science. Though this thought cannot be adequately developed here, I would like to suggest what it might mean. To do so, however, we must take a moment to reflect on the notion of "dialectic" itself. For though I have often referred to the concept in the course of this study, and made certain

---

[31] In connection with this account of dialectic, the reader might like to compare the discussion in the article Dialectic in the *Syntopicon* (*GBWW*, Vol. 2, pp. 345ff.), where somewhat different "benchmarks" are employed.

claims about it in passing, I have not explained what I understand it to mean.

Dialectic has had its own history in our Western experience; we might speak of its own "dialectical" development—or of "the dialectic of the dialectic." It will suffice for the moment if we take just three great benchmarks of dialectic as exemplars of this unfolding of the concept: the Platonic dialogues, of course; the Hegelian dialectic; and the dialectic of *Capital*.

Socrates remains the model of all dialectical teaching, we might agree—but how do we understand this? Perhaps, simply, that he teaches by means of *real questions*. Life stands or falls by the answer to a Socratic question, and the answer is always entirely up to us. It is the conviction underlying all dialectic, in any of its forms, that such fundamental questions of value, right, and human purpose—questions that have no "objective" answers, but that touch on matters closest to our lives—are not idle, but that in one way or another, we do have access to crucial means of moving toward their resolution.

On the other hand, the term "dialectic" suggests also a certain structure of inquiry, and we need to consider the relation between such a real question on the one hand, and the pattern of dialectic on the other: not only the vividness of the human questions that they ask, but a certain common form links Plato, Hegel, and Marx. As we say this, however, it is very important to avoid possible misunderstanding: though the thread that links these three stages of the dialectic is very real, and fundamental in particular to our understanding of Marx, to point to this common principle is not at all to assert that dialectic is not very much transformed in its passage from one stage to the next.

The dialogue *Meno* serves well as a paradigm of the dialectical motion in its Socratic mode, and I would invite the reader to reflect with me on it in relation to the very brief remarks which follow (*Dialogues of Plato*, tr. Jowett, Vol. I, pp. 349ff.). In broad terms, we can see that it begins, as it should, with a question. It is not easy to say just what that opening question *is*, for as Meno asks it, it contains a nested set of complications that may well be endless. "Do you," he asks Socrates, "have it in you to say to me whether virtue can be taught, or if it is not teachable, whether it comes by practice; or if it neither comes by practice nor can be learned, whether it comes to men by nature or in some other way?"[32]

I think it is characteristic of dialectic that the question itself is already questionable: we have questions in echelons, questions about the question. Is this a question about Socrates (as it seems grammatically to be), and his power, or is it about virtue, or about teaching—or about Meno himself? All these elements are so problematically interrelated that we sense that they must be, finally, just one question: who teaches, who learns, what is taught, what teaching is, and where the answers come from—a single package of perplexities.

Marx, similarly, asks us, as participants in the woven world of capitalism, to wonder with him who we are, what virtues we practice, and what source of light there might be for us from somewhere beyond this system within which we are enclosed, by which we might judge ourselves and it. Finding the real question about capitalism has been our first problem in approaching this attempt at a rereading: it is not, I have claimed from the outset, a book "about economics." But it is not so clear, on the other hand, where the boundaries of Marx's real subject matter do lie—his questions penetrate, as those of Socrates do.

To focus, then, for a moment on the *Meno* as paradigm of the dialectical form: the opening question is developed in a variety of artful ways that steadily reduce this brash young general, initially confident enough in the world's ways, to a state of what may be serious wonder and concern. He meets unexpected difficulty in defining virtue ("excellence"), though he had evidently never before doubted that he was himself a living

---

[32] (*Meno*, Steph. 70) I have taken the liberty of supplying my own translation, to catch what I think is the literal, or sub-literal, intention of the Greek.

model of it. His failure leads him to a certain, perhaps petulant, despair, and it is significant that here the dialogue comes to a dead stop—at its effective center—with an outcry from Meno that seems in its own terms unanswerable (*Meno,* Steph. 80). Dialectic is impossible, he asserts, because we either (1) know the thing we are looking for already, in which case it is idle to be searching, or (2) we do not know what we are looking for, in which case it is even more idle to be searching, since we wouldn't know it if we found it. This is, presumably, a standard sophistic argument against learning and truth, but it defines a real problem, and here it may be that Meno is genuinely struck by it. Meno blames this *impasse* on Socrates, who he says is like the electric torpedo fish, which shocks anyone who comes upon it into numbness. This point of death of the argument— which appears, I am tempted to say, at the virtual center of all dialectic—is often referred to in Greek as the *aporia*—the sticking point, the point of no passage, the point of no return for the argument.

We cannot here trace that way by which Socrates at once opens a path for Meno—and, we must add, for us. It is by means of "recollection," Socrates half-mythically explains, that we are able to assert truths with conviction, as if we once knew them and were recovering them through a mist of forgetfulness. The *aporia* of the dialectic will always lead us into the darkest obscurity; but if all goes well, we will emerge empowered with knowledge we had not known we possessed. To be schematic about this, and as an aid in tracing something of this same pattern in Hegel and Marx, we may say that there are three parts to every dialectical motion.

I.   The opening question, a real question, which takes form through an intense searching in the mode of questions and answers, not yet fully articulate;

II. The clarifying argument to the point of contradiction and despair; the question becomes articulate but, at the same time, leads to *aporia*;

III. The passage beyond the *aporia,* through yet more serious questioning, which yields whatever knowledge is humanly possible; not, however, in the mode of syllogistic consequence, but drawing on some larger source of human intuition, in the form of image, myth and mystery.

If we were in this reminded of other tragic trilogies, such as the *Oresteia* of Aeschylus and the Oedipus cycle of Sophocles, we would surely be on the right track. [33]

Now, Hegel. The pattern of dialectic is in many ways the same, but there is surely a fundamental difference as well, for we are in a different world: the Judeo-Christian world, the world of the omnipotent, omniscient Creator God, God of love and sacrifice—the world infused with the Holy Spirit. It would be hard to think of any aspect of life that was not touched by this world change. The Hegelian dialectic thus takes as paradigm not only Plato but the Christ story: Advent, Passion, Resurrection. The dark moment of the *aporia* becomes the suffering on the cross. Dialectic is the journey, not simply of the learning mind, but of the subjective Self as Spirit, creatively acting to shape its own image, and passing ultimately to freedom—not in objective knowledge—but in self-consciousness. The argument unfolds, not in an afternoon, but through the ages of history. The course of history now *is* the course of the argument.

Hegel puts this in a pattern temptingly like the schema we outlined earlier, but deeply altered as well:

---

[33] We may be reminded as well of the dark moment in the *Phaedo,* the dialogue in which we witness the last hours, and the death, of Socrates. At a certain point, the argument, and with it, possibly, all the powers of dialectic to lead us to the truth, seem to have failed. A silence falls over the group at this apparent death of the argument, prefiguring Socrates' own death (*Phaedo,* Steph. 88–89). In the mythology of the dialogue, this dark moment is the depth of the labyrinth of Crete. The rescuing dialectic of the third phase becomes the thread of Ariadne, by which Theseus is led back to the light.

I.   The Spirit *an-sich* (in itself, immanent), thrusting forward
     in dialectical search for its own identity;

II. The Spirit *für-sich* ("for" itself), objectively determinate
     in the world, and suffering all the consequences of the
     estrangement from itself;

III. The Spirit *an-und-für-sich;* the mystery of the Resurrection,
     which restores the Spirit to itself without denial of its
     objectification.

The third stage is expressed in a German word of double
meaning, *Aufhebung,* at once both "cancellation" and "uplifting."
"Transcendence" is a word that is often invoked by translators to
suggest something of the magic of this third phase.

It would be worth the reader's turning to a paragraph in
the *Philosophy of History* in which Hegel images this life of the
Spirit in terms of the passage of the sun in the course of one
great day of the history of the Spirit in the Western world (tr.
J. Sibree, p. 103). The first phase (or "moment," in Hegel's terminology) is that of the dawn, but no ordinary dawn: to suggest
the sense of that immanence (the Spirit *an-sich*), Hegel visualizes the experience of a blind man who has for the first time
been granted sight (I). He is engulfed in the light of the rising
sun: no objects are yet differentiated; it is a single, whole experience. By midday (II), however, objects are differentiated, the
magic is dispelled, and all is explicit and objective; the Spirit
is alienated, distanced from itself, *für-sich.* In the life of mind,
this is the phase of objective, syllogistic reasoning, which Hegel
interestingly likens to attaching labels to a skeleton. Yet by evening (III), Hegel says, man "has erected a building constructed
from his own inner sun" (ibid.), a vision of his own Spirit, and
in this late light, Spirit knows itself in the illumination of its
own, greater sun. This is *self-consciousness,* and that, for Hegel,
is freedom, the culmination of the dialectic.

The dialogue which, for Plato, has been in a sense timeless, essentially the same in every repetition, has now entered
time and become History itself, one epoch perishing, to be

transcended by another in the advance of an argument that has become our collective inquiry—the learning process of Western man. If there is indeed real *progress* in the evolution of mankind and the development of our human culture, must it not be the case that Hegel is right, in some fundamental way?

Finally, Marx—and the bearing of this long tale on the dialectical reading of *Capital*. It seems to me that without having reached some understanding together of this dialectical tradition, we could not recognize the nature and magnitude of the task Marx may be setting for us, his readers. Commentators often speak of Marx as having rejected Hegel's thought, while using his "method." It is true that Marx does explicitly reject Hegel's philosophy in certain important ways, while incorporating the form of the dialectic. But dialectic is never simply a "method." Such a tortured procedure, through the negation which marks the second phase, makes sense only if the human situation is itself perceived as dialectical, only if the human circumstance demands and justifies such a process—as to which all three of our "dialectical" authors must be in some deep agreement. What is it about our world that leads Marx to conceive it—as did Plato and Hegel before him—as a dialectical problem, and in what way, in *Capital,* do we meet the threefold structure of the dialectical investigation?

Marx has found that capitalism is a social structure in motion, in history, generating and regenerating its own progression, and changing as it goes. We, his readers and fellow human beings, are, like Marx himself, immersed in that world and in the stream of that unfolding history. Its terms and symbols are ours. Yet at the same time, it is eminently questionable: *what is it,* and hence also, *who are we,* as participants—our selves and our lives embedded in it? This last is certainly not an idle question, certainly not an "objective" one, but a truly Socratic question; it is our subjective selves that Marx places in doubt. The dialectical question is always one of life and death. *Capital* is such a dialogue, with the opening question addressed to us. [34]

Perhaps Aristotle has asked the opening question in its root form: what is the relation between *economics* and *chrematistic?*—that is, in terms of our own time, what is the relation between human ends (use-values) and the system of capitalism (the universe of exchange values)? From the immediacy of the concrete economic life that is so familiar to us that we cannot criticize or even really see it—phase I of Hegel's schema—*Capital* has proceeded to separate and trace the abstract forms that are in fact at work. Marx's formal theory serves to reveal the systematic network of abstractions that constitutes the system of capitalism; we recognize the objective abstractions of Hegel's phase II in the reified entities of the bourgeois trinity. As we penetrated deeper and deeper into this labyrinth, two things were happening: exchange value as an abstraction from all judgment of human value became total, and at the same time a system of illusion was shaping itself in such a way that all traces of the derivation of the system from its human base were swallowed up. That is, for Marx, the depth of the labyrinth—we have ceased at this point even to ask the question of human goals, have ceased even to recognize how far we have strayed from them. That is where he finds us today, and in the three volumes of *Capital* he has taken the full measure of our darkness: the depth of the *aporia*. Our systems and our machines, including those of war, dictate to us: we can no longer bend them to our human purposes. It is at this point that Marx utters the sober words: "It might seem that we must abandon all hope…." [35]

---

[34] The opening question of the dialogue *Gorgias* is perhaps really the essential opening question of all dialectic. Socrates suggests that they ask of Gorgias, an eminently successful and celebrated orator, "Who he is?" (*hostis estin*) (Steph. 447). This becomes the occasion of a tragic fall, in which he is revealed—to himself as well as the assembled company—as having led an unjust life.

[35] *Capital*, Vol. III, p. 252.

What light is there which will illuminate our present darkness?

If Socrates invokes the Forms, and Hegel invokes the Spirit, what power can Marx turn to that will bear the dialectical burden of a resurrection of the human, in some new form, in the midst of a world in which quantity has seemingly so totally devoured quality? I think in fact we have seen the answer unfolding as the account of capitalism progressed. In the *Meno,* we sensed the operation of the mystery of "Recollection"—that is, the mystic advent of knowledge from a source that was in no way evident—as the dialogue unfolded. So in *Capital,* as the picture of the system of capital was drawn more and more completely, Marx has been continually indicating that something else is happening, not evident on the surface of the system itself. This is, as he has traced in concrete and immaterial terms, the increasing socialization of the processes, the socialization of labor, the structures of credit and the markets, all the devices of science and engineering, of communication and distribution. It all derives, to appearances, as a weaving of mere impulses toward private gain. But despite such apparent competition and separation, we are as human beings in fact learning, albeit in some sense despite ourselves, ever more ways to function cooperatively and socially.

Like any answer to a genuine dialectical question, this new society of rational and cooperative effort is too urgent for us to perceive neutrally, as mere *observers*; rather, it appears as an answer to alienation, which has become an anguish for us in our time. A dialectical question takes the shape of a crisis which we suffer (Hegel's phase I); to ask the question is to cry out for release from a bondage.

What Marx is trying to show us, as the dialectical subtext to our culture of competition, automation, and strife, is a possible answer to the longing of the denial that is our modern world. Out of the depth of the negation (phase II) arises a new prospect (phase III): not simply as a denial of the alienation, a negation of the negation, but as a positive affirmation of an access to a human community that had never existed before.[36] I have suggested that the question which we launch returns to us: we are set to wonder about our own identity. Could we be members of such a human community? I think that *Capital* is leading us in the direction of what might be a surprising answer: we are rational and social human beings in ways we had perhaps thought impossible.

This same pattern can be seen, arguably, in terms of a long motion of human history, one that I think is the underlying clue to the meaning of *Capital* for Marx. In Athens (I), the *polis* was a sketch of a human community: not yet recognizing human rights, human equality, the full worth of the individual human being. It was still a slave society—yet, nonetheless, a society that saw itself in human terms, shaped to human purpose and the human good. Quantity was still instrumental to quality.

We have now in capitalism the full development of the opposite (II): the farthest extreme of departure from the human and the qualitative, and from the sense of human community—the *aporia* of our Western history. Yet even here something is brewing which is new on the face of the earth, a new sense of our common humanity, and of a new human society which would give reality to it.

-----

[36]It seems likely that the threefold pattern of dialectic is indeed reflected (or founded) in the traditional structure of the Greek tragedies in the form of trilogy. Thus Aeschylus's trilogy, the *Oresteia,* opens with what might seem the universal outcry of suffering mankind: "I pray the gods for release from these labors…" (*Agamemnon,* I, 1). Only after the depths have been sounded of the tragic events of the first and second plays does the "release" in fact come, in the form of the transformation of the Furies into benign spirits, and the foundation of the Athenian polity in a system of justice. These identities are suggested in the closing scene of the *Symposium* (Steph. 222–223).

This would be the *Aufhebung* (III): the cancellation of the present alienation, and the affirmation of something that we cannot yet know in detail or clarity but can begin to perceive: a rational human society which Marx calls, tentatively, not "political," but an "association." In it would be realized the value and separate identity of each individual, whose formal recognition has been the triumph of our own political heritage. But this equality and these freedoms would be brought into actuality by recovering the sense of community that was the legacy of Athens, and which we have so nearly lost.

Marx sees, then, first, the affirmation of the human community in the *polis,* where individual rights and the equality of all human beings were not yet known; second, the recognition of these rights and the formal commitment to human freedom in our own society, where on the other hand we have gained rights only by denying the human community— we have gained formal freedom only by separating ourselves from one another and substituting quantitative processes for a common rational judgment of our human purpose. The third phase of our Western history is already with us, but we have not seen it: the incorporation of the community of the *polis* with the technology of the modern world, in the realization of that individual freedom in which we so strongly believe, but which our present society still contradicts at every turn. True substantive individual freedom is now a realistic possibility. This, I think, is the constant suggestion of the pages of *Capital.*

Dialectical Science

Is Marx, we must ask, writing *science,* or is he painting yet one more dream-picture of a future for mankind? The question would hardly arise, if it were not for Marx's surprising claim that this is indeed a work of science, and by no means mere speculation or political persuasion.[37]

Evidently, Marx is challenging our idea of "science" in a way that requires us to go back to its foundations if we are to follow his thinking. Such an investigation, of the possibility of a *dialectical* paradigm of science itself, would be altogether beyond the compass of our present study. Yet if Marx is claiming that *Capital* in its dialectical aspect is at the same time to be understood seriously as a work of *science*, then we can hardly omit consideration of this claim altogether from our own "reading" of the book.

I think it will be possible here to take just the briefest measure of what Marx's claim concerning the scientific character of the dialectic—or the dialectical character of the sciences— might entail. It may be that if there is indeed a formal, Newtonian theory of capitalism housed in *Capital* within a dialectical framework, as we have discussed, the scientific character of the work belongs more to the dialectical frame than to that objective, formal theory. Can a work be at the same time "dialectical" and "scientific"?

There is an image of "good" science, shared by both our Ptolemaic and our Newtonian models, for which the criteria run something like this: the theory is based on explicit assumptions expressed in univocal language; the reasoning from first principles is sound; the conclusions reached are capable of being tested in the observatory or the laboratory and are thus confirmed or disconfirmed by observation or experiment. Nature renders a dispassionate judgment; and a theory of this kind can be regarded as objectively true just insofar as

---

[37] The claim is made, for example, at the outset, in the Preface to the First German Edition: "…it is the ultimate aim of this work to lay bare the economic law of motion of modern society…. My standpoint, from which the evolution of the economic formation of society is viewed as a process of natural history…. " (*Capital*, Vol. I, pp. 20–21)

experiments have been devised and carried out to put it to empirical test.

I am oversimplifying, of course, but in broad terms this is the paradigm that accounts, we tend to feel, for the brilliant success of the sciences in the past three centuries. As theories fall and are replaced, with what seems increasing rapidity, we become more sophisticated in our recognition of the fallibility of any one theory, and yet the paradigm of "science" itself remains firm among us: in fact, the more fallible the individual theory, the more crucial the adherence to strict method in the overall community of the sciences might well seem.

What *was* that idea, which was born into the world in the course of those events that we call the "Scientific Revolution"? The crux, I think, is the concept of "objectivity," which has already appeared so often in this study. It is a notion which seems so persuasive to us that we embrace it in all aspects of our society, whether in our journalism or our personal lives, with nearly the same conviction with which we insist on it in our laboratories.

Let me for a moment, for the sake of reflection, paint the alternative picture. Perhaps the world does not really come apart, as the criterion of "objectivity" presupposes, into two distinct parts—the realm of the observed and the realm of the observer. Perhaps the "subjective" and the "objective" are inherently and inseparably joined in a single fabric, and perhaps this fabric is itself the work of time, the weave of history. When we enter the laboratory and arrange an experiment, perhaps we are fashioning images of ourselves and our historically conditioned expectations—in the forms of the apparatus, shaped by the machines of our time from the materials we have wrought from the earth and cooked up in our processes of production; in the devices of measurement, fitted to notions of time and distance and all the entities of our current speculations—indeed, to our paradigmatic vision of the nature of "theory," or of "knowledge" itself. Marx says, "We hear with a human ear, and see with a human eye."[38]

Our very senses and the entities they perceive are themselves products of human history. They pick out, form up, and thus "detect" what we have learned to sort, attend to, and speak of. These comments are obvious enough, perhaps banal, but their consequences for the idea of "science" may be profound if we take them altogether seriously, as it appears that Marx does. What we "measure," "observe," "record," and "prove" or "disprove" in the laboratory belongs inextricably to that same web of human history in which we ourselves are involved— when we seek the object, we, to a large extent, find the subject. When we turn to Nature for final judgment, we meet, often enough—ourselves. "Man makes Nature," Marx says, meaning, I think, that everywhere we turn, we meet ourselves in domains of our own shaping and making.[39]

---

[38] "The eye has become a *human* eye when its *object* has become a *human,* social object, created by man and destined for him. The senses have, therefore, become directly theoreticians in practice…. It is evident that the human eye appreciates things in a different way from the crude, non-human eye, the human *ear* differently from the crude ear." ("Private Property and Communism," in Bottomore, ed., *Karl Marx: Early Writings,* p. 160.)

[39] Marx says of the naïve materialist: "He does not see how the sensuous world around him is, not a thing given direct from all eternity, remaining ever the same, but the product of industry and of the state of society; and, indeed, in the sense that it is an historical product, the result of the activity of a whole succession of generations, each standing on the shoulders of the preceding one…. Each of the objects of the simplest 'sensuous certainty' are only given him through social development, industry and commercial intercourse." "Feuerbach … mentions secrets which are disclosed only to the eye of the physicist and chemist; but where would natural science be without industry and commerce? Even this 'pure' natural science is provided with an aim, as with its material, only through trade and industry, through the sensuous activity of men." *(German Ideology,* pp. 62–63.)

Of course, there is a vast component of the "objective" in the work of the laboratory; such "objectivity," we tend to say, "built the atomic bomb." Yet we must check ourselves; we know better. "Objectivity" alone did not build the bomb, nor the deadly train of terrors which continue to follow in the path it opened. What we call "objectivity" melds with the social and the human in ways we have not very well learned to disentangle. The bomb was the combined product of inextricably interrelated "objective" and "subjective" factors. What we think of as "objective" science is, perhaps, *inherently* "framed" within a context of larger social institutions and the movements of human history, much as I have claimed the formal theory of *Capital* is "framed" within a dialectical critique.

It is not hard to see that our image of "objective" science is very closely wedded to the idea of capitalism: scientific truth, we learn to believe, is quantitative, and objective realism is identified with power: the structure of truth, we believe, is hierarchical. Any other forms of truth—that is, most of the truths which really matter to us, humanly speaking—tend to be misprized among us as matters of "opinion," or (worse) as "value judgments."

Our very idea of "objective" science is a product of the dialectic of history, of course: it has emerged from a certain human experience, a certain human preoccupation, and it will pass, as experiences broaden and we gain fuller recognition of what a "human" preoccupation might really be. Dialectical science, which looks at the rise and transformations of "objective" science, takes a much larger field of view. It sees all that "objective" science sees, but it sees as well the extent to which so-called objective science is the dialectical work of time: how its preoccupations, its univocal terms, its underlying ideas flow, shift, and are transformed; how science takes the image of the social processes of an age: how it serves them, is rewarded by them, and is in time transcended. Because "objective" science does *not* see all this; because it takes its terms as "univocal," because it thinks its intentions are indifferent to the biases of institutions and purposes, which in fact fund and house and nourish it, because it does not see the colorations of the minds and passions which give it being, it is not "objective" at all, but in its presumed objectivity, to that extent, merely naïve.

The picture cannot see the frame. Only the whole view would take the measure of truth. Marx strives in *Capital* for this wholeness of view: that is, for the more complete science. And we might reflect that if, with respect to the political and the social, he is showing us the way to a recovery of a trans-formed ancient *polis* in the modern world, so with respect to the concept of science, he is showing us the way to a recovery of the principle of *dialectic* in relation to our modern truths. It must be important to observe that the two—the social and the scientific—are strictly parallel enterprises. If Marx is right, our limited idea of "science" and our limited idea of "society" mir-ror each other, in their pallor.

The customary term "dialectical materialism" seems to serve well to express the union in *Capital* of the ancient dialectical and the modern scientific traditions. It is a work of science, but it is not, on the other hand, an "objective" work. If it were objective, if it were merely examining the *fact* of capitalism "sci-entifically," in the manner of the formal, Newtonian theory, it would *not* be scientific. It is in fact a human work in which alienation appears as an affront, outrage surges in its prose, and the prospect of something coming to birth appears with the urgency of prophecy. What do we want to "know," and what does "learning" feel like? Dialectical materialism recognizes that our concern is with the forging, in history, of our own humanity: we have not yet learned what it is to be human, and that must be the governing question for our science. It is not as though we had the option of setting it aside—bracketing it, for separate consideration. Without our recognizing it, this one question of ourselves infuses and informs all our sciences.

The question of our identity is urgent upon us. Though we no longer eat each other, we currently tear ourselves limb from limb—men, women, and children—with daily and procedural indifference, by means of our "objective" sciences, and in the name of "freedom," a word whose meaning we evidently do not yet quite understand. It is in the process of working our way out of this dark complex of circumstances that we must come to find ourselves: the "knowing" is not simply a "theoretical" matter, but rather a question of shaping our practices—above all, surely, our practice of the sciences—little by little confirming our horror at our mistakes, little by little finding ways to do better. Learning in this unfolding, dialectical practice is what Marx in some places calls *praxis*: not simply theory or simply practice, not simply "objective" nor simply "subjective," but a process of bringing our humanity into being in and through conscious practice.[40]

The emphasis in dialectical "materialism" is on the principle that we can come to grips with this only in the detail: the detail of the conditions under which our world lives—the society, the language, the physical equipment, the myths, the arts, the health or disease, nourishment or starvation, peace or war, wealth or poverty, which are the "material" of our actual lives. "Learning" is not a question of general ideas, simply, but of the detail and suffering of human *praxis*, in, through, and beyond the concepts, the theories, the equations, and the terrible misunderstandings and consequences in which our "sciences" are entwined.

Conclusion

I suggested at the outset that *Capital* was not a work on economics but was really directed at other and larger questions. I think we have seen that Marx has really laid his question at

---

[40] Marx sets forth the notion of *praxis* in the *Theses on Feuerbach*. For example, from the Second Thesis: "The question whether objective truth can be attributed to human thinking is not a question of theory but is a *practical question*. Man must prove the truth, i.e. the reality and power, the this-sidedness of his thinking in *praxis....*" (*German Ideology*, p. 121.)

our own doorstep and challenges our concept of ourselves. Are we the separate, competitive beings envisioned by Hobbes and Locke, and presupposed by Adam Smith, or are we in some fundamental way members of a larger society? Aristotle thought we were political "by nature"; our current view is that the polity is a work of our own making, and that it serves to bring together persons who are initially and primarily alone. Aristotle thought that outside of society we were either beasts or gods; our mythology has it that we were free in our separation, and that we form and join societies in order to preserve that original freedom. I do not think we have really resolved this question. There is, I think, a deeply felt conviction that something is missing in our modern societies, and it may be the sense of society itself. It would not be a particularly original observation to remark that we moderns feel lost and insecure; we feel a need to "belong" but are not sure what we are prepared to belong to. We try to fill this gap in religious communities and in networks and organizations of all kinds. They express something we are aware we are seeking, but they are all filling in, in their various ways, for a missing center.

Marx, I think, is suggesting that we have misplaced our conception of freedom. We do not become free, he is saying, in isolation: rather, we can only be free in common. All our recent intuitions warn us against this approach: we have seen too vividly the consequences of totalitarian social impositions of "freedoms" that convert to nightmares. But *Capital* is not advocating totalitarian solutions: Marx's vision—right or wrong, possible or impossible—is of a creative individual freedom in practice, achieved within and through membership in a conscious and rational society.

Marx is, perhaps, too much of a democrat. He tries hard and in detail, even from his vantage point of a hundred years ago, to show us that we are already doing many of these things: we do have these capabilities, we really do think together and plan, we can formulate a common end, can amass great social powers and inventiveness, determination, and skill to accomplish social goals, and can in the process achieve new levels of individual freedom in practice. Perhaps in all this he remains unpersuasive; the cautious reaction is to point out that something called "human nature" will not admit cooperation or common reasoning to be a human good. We will, caution pronounces, pursue private ends, tear and maul one another, or lose all interest if we can see no separate gain. Perhaps in this we do injustice to ourselves.

One word has been strikingly absent from this discussion—the word *revolution*. It is not a word which comes up much in the text of *Capital*. Surely *Capital* is a revolutionary text, as all dialectic is; and it is clear, too, that *Capital* is revolutionary in a way that Plato's *Dialogues* are not (though it is a curious thought, and a testament to our English freedoms, that Socrates was executed as a threat to the polity, while Marx died a natural death as a father and a husband). The difference, I think, is that while Socrates mortally affronts Athens with his insistence that it has its values upside down, he sees this as a tragic circle out of which, in the long run, no society can hope to escape, and his arguments therefore do not suggest revolutionary change. *Capital* on the other hand is implicitly revolutionary, because Marx says we are stronger and better than we know, and we have it in us to constitute a society that affirms our humanity in ways we have not yet realized. In this sense Socrates belongs to the tragic tradition, and Marx to the prophetic.

Throughout the world, for a century now, people have been reading *Capital* as a message of hope. The spectrum of interpretations has ranged from the most esoteric formulations of "the dialectic" to discussions of revolutionary block committees dedicated to life-and-death struggle against the oppressions of the modern world. I am suggesting that we cannot afford to remain illiterate in relation to this world discourse. I urge readers of the *Great Books of the Western World* to take down Volume 50 from its very likely neglected place on the shelf and see what they can make of this remarkable package of propositions about ourselves.

Afterword

Afterword

I HAVE CLAIMED that mankind today is held in a form of bondage, the essence of which is a paralysis—partly intellectual, but to a large extent social or habitual—that blocks us from bringing the best powers of our minds to bear on the problems we care most about.

We know, for example, how to bring the most advanced methods of the biological sciences to bear on the problem of increasing agricultural productivity—yet we fail utterly even to formulate, much less solve, the problem of distributing the resulting crops to feed the starving populations of the world. The first is recognized as a *scientific* problem; the other, we somehow imagine, is not. Yet what indeed enables us to solve the one problem, yet blocks us from even approaching the other? Mind, I conclude, is in bondage to forces which, admitting the methods of science into a vast array of fields, arbitrarily exclude them from other areas of even greater concern—wantonly leaving those areas to the obscure forces of opinion, prejudice, strife, and, often enough, warfare and destruction.

There is no need in principle, I am sure, to accept such bondage, and I have summoned the aid of our three authors in making my case. They open a much clearer vision than we normally get of the true powers of science, and they argue in very different ways for its universality. They offer the prospect of a release from the limitations in which intelligent minds today everywhere find themselves held. I am not, certainly, conjuring the image of a panacea but rather suggesting a new resolve—and a source of courage in the examples of the challenges against which these three minds themselves struggled.

They all three show us how universal the sense of struggle has been; we are not alone in our sense of subjugation to an opposing force. We have traced this development of new thought in struggle with negation and denial back to Plato. We identified it as *dialectical*—first in respect of overcoming obstacles, but chiefly in confronting stark contradiction, out of which discoveries emerge as radically new. In this way Newton's battle against Descartes's reduction of the world to a mathematical machine gave birth to an altogether new vision of a world that was both mathematical and infused with spirit throughout. Maxwell, in turn, battled what had become an ensconced autocracy of the physics of Newtonian action-at-a-distance to produce an entirely new, and newly democratic, physics of the *field*. Finally, Marx, bringing the new science to bear on the system of capitalism itself, showed how capitalism, though it inherently alienates human values, produces the beginnings of a new society that is truly human and truly free—one that would incorporate and transform all the advances capitalism itself had produced.

The life of creative thought, we may conclude, lies in a tension of contradiction. Further, we have seen that these struggles are not isolated, but build upon their predecessors. No earlier step is lost; but as the dialectic enters history, the human spirit in a very real sense links eras and grows from generation to generation. We draw today upon the powers that Descartes, Newton, Maxwell and Marx have built into the body of the liberal arts as we know and use them today. We may well take heart from our review of their works to believe that our struggle, however unresolved we ourselves may leave it, will contribute to an evolving structure of free human thought that belongs to a prospect beyond the horizon of our vision today.

We might ask ourselves, in review, what we most prize as our inheritance from each of these authors—while acknowledging that each of these inheritances must be earned anew by a reading that looks beyond the conventional interpretations.

Newton, when read in the terms in which I believe he wrote—so different from those in which he is usually remembered—has two specific powers to grant us. The first we might call the gift of mathematics in its role as a rich resource for every searching human mind.

Most people today regard mathematics as an instrument of calculation useful to specialists of one sort or another, primarily engineers or scientists. Few in the course of their educations have met mathematics as a fascinating and beautiful resource for broader human thought to which everyone is entitled. But when we surrender mathematics to the specialists, we gratuitously abandon a resource we need and might well delight in, and we commit our own minds to limitation. Newton, as we saw, was a strong opponent of the reduction of mathematics to a mere algebra: he preferred, and used to the fullest, the methods of geometry, which he artfully shaped to bear rich symbolic value. In the same spirit he rejected the reduction of nature to an algebraic, mindless machine, and formulated instead a philosophy of all nature that was at once fully mathematical and yet in no way reductive.

What we may take from Newton, then, is a resolution, first, to restore mathematics to the range of arts available to the liberally educated mind—and, further, to employ mathematics freely in the solution of human problems without being intimidated by the argument that their use would be reductive. Merely mechanical application of mathematics is indeed reductive—not, however, because it is mathematical, but because it is mechanical. Creative use of mathematics as an instrument of the free mind need yield to no such criticism. Only entrenched social prejudice and bad educational habits clip the wings of the free mind by denying it the tools that actually lie ready for use, and are so urgently needed today.

Newton's second, still larger legacy is ours if we are wise enough to claim it. He perceived biology, theology, chronology, alchemy as belonging to one body of coherent thought. We need to take very seriously his dedication to that vision of wholeness, and not ourselves flinch from seeking such unity in our own time. For it is only in relation to a *whole* that any part can have meaning, or can promise truth beyond convenience. We have seen how for Plato this grasp of wholeness characterizes the faculty of reason, *nous,* which stands highest on the Divided Line. Plato calls such search *dialectical* and, in his terms, we may say that "science" is not truly science if it is not dialectical—that is, if it is not grounded in an ongoing search for the surest attainable foundations of truth.

Our primary insight is that humanity is one. Despite the splits and varieties so apparent among mankind's myriad beliefs, the aim of minds everywhere is to arrive at community of understanding. Minds everywhere seek conversation, the token of membership in the human family. Religious differences stand as welcome monuments to human diversity, but beyond that, we strive to capture in speech those principles that ring true in every language, and that bespeak one world, just, productive, and at peace.

Newton carries us very far, opening the way to a creative use of mathematics and mathematical sciences. Yet Newton's thinking is still in a sense feudal and autocratic; as we have seen, he thinks in terms of lordship and obedience. By contrast, human reason is inherently *democratic,* everywhere rebelling at the imposition of unexplained laws, and *dialectical,* relentlessly questioning authority and demanding explanations. Maxwell rebels against entrenched authority, both within physical theory itself, and in society and its educational system, where by Maxwell's time mastery of Newtonian methods had become essentially a privilege of class and a weapon of social autocracy. Field theory is the democratic revolution within the sciences, carried out in accord with reason's demands that gaps be filled and wholeness be restored, both within the sciences themselves and in the body of society.

Where the physics of forces begins with individuals and builds the whole, as it may be, by sheer aggregation, field theory immediately grasps the whole as primary. It derives particular actions, possibly in more generalized, less literal terms, as organic constituents of this intelligible whole. It is striking that this whole, which can be visualized directly in terms of geometrical patterns such as Faraday's beloved lines of force, is accessible to mind with a minimum of formal preparation. To Maxwell this is an insight into the nature of mathematics itself, and he is in full earnest when he claims that Faraday, with his magnetic patterns, is the real mathematician of them all. The highest reason is not interested in long trains of arguments as such, but only in the insights which may emerge from them. Whether in physics or in any other area of its interest, reason—in the democratic form Maxwell gives it— makes what is effectively Faraday's demand: wrap truth in the simplest, most revealing form, and share it with all on behalf of mankind. So where Newton saw his wondrous mathematics as a secret instrument for the use of select, wise rulers, Maxwell and Faraday transform this mathematics into a new instrument, the *field*, accessible to all, and best adapted to meet the demands of the democratic intellect.

Essentially these are the demands reason has always made— even of the Platonic *nous*—for a grasp of that unity which gives meaning and intelligibility to the parts, and which sees the whole as meaningful, beautiful and good. Maxwell may be challenging us to carry on this democratic revolution in our own time by engaging the powers of reason to address real problems of greatest human concern. Field theory is just one model of the way in which this can be done; it points the way, however, to relief from the bondage of ignorance and denial we have seen imaged in Plato's cave.

Our struggle today is to conceive our own planet, with its many converging and interwoven systems, as a *whole* on the model of the *field*—and then, armed with this insight, to preserve it from the destructive effects of competition, strife and greed. For this, we will need to corral all the powers that mind possesses, most especially those which Maxwell is opening to us in his *Treatise*.

If we question, as many surely do, the possibility of applying to social systems these methods which have proved so successful in the modern sciences, Marx shows in one stroke that we need have no such doubts. In *Capital* Marx parallels Newton's account of the system of the heavens with his own closely corresponding account to show how capitalism, an equally law-based system, actually works. Since, however, his analysis inevitably at the same time reveals the inherent injustice of capitalism, it is by its nature a revolutionary text and has been unwelcome, and its central theory largely ignored, in much of Western thought and education. What might otherwise have become a major new source of strength for the democratic mind has been effectively locked away, its denial one more element of the bondage in which free thought is held.

Marx sees, far more vividly than most readers do today, the nature and scope of a trap that constitutes the all-enveloping cave of our present time. The obscurity of this modern cave is so familiar and so apparently total that very few dwellers in this bondage can even catch Marx's meaning. Marx's point is that to *alienate* one's own labor, mental or physical, is not freedom but its opposite, servitude. Most of these cave-dwellers will not understand Marx when he calls this process *wage-slavery* and suggests that its remuneration, even when abundant, forges only *golden chains.*

This is the challenge which free mind meets when it attempts to turn the powers of scientific thought upon this very system, which surrounds it on every side. Science—so the cave demands—is to be directed only to matters which advance the system of shadows. It is to be applied to making capital grow, not to the solution of real human problems like the distribution of food, health, or knowledge. Any extension of serious, scientific and intelligent thought beyond the bounds to which it is normally confined raises the dangerous possibility of criticism of the encompassing system itself. At this point we have come to the heart of the question of bondage with which our Preface began.

Newton and Maxwell overcame huge barriers to open new powers of scientific thought in the service of the free human mind. With the further help of Marx, perhaps we now see better why our culture teaches us to think of these powers as being only of limited interest—that they are for the use of specialists confined to well-defined domains of technology and science.

Marx shows with devastating clarity that if the light of intelligent, scientific thought is thrown brightly on the walls of the cave itself, the true nature of the system and the fact of bondage will be starkly revealed. As in earlier eras, the free human spirit today cannot but rebel against such bondage once it is detected, and then work to find some way forward which searches out all that is best from the past, indeed, but opens the prospect of some new and better way in the future.

It seems to me now that I must have first written, and then collected, these essays with some such intention latent in mind, sensing the importance of opening Newton and Maxwell to fresh reading, and hoping that Marx might cast such a light that we could no longer endure to live at ease in this modern cave, in which, as it seems to me, he so unmistakably reveals us to be enclosed. If such is indeed the case, might we not agree with Marx when he says that the point is not merely to know the world, but to change it?

Bibliography

Aeschylus: Plays, tr. G. M. Cookson (London: Dent, 1956)

Archimedes. *The Works of Archimedes* with *The Method of Archimedes*, tr. Thomas L. Heath (Dover Publications, 1950)

Aristotle. *Basic Works of Aristotle*, ed. Richard McKeon (New York: Random House, 1941)

Bacon, Francis. *The New Organon and Other Writings* (Library of Liberal Arts, 1960)

Barnett, S. J., "A New Electron–inertia Effect …," *Philosophical Magazine* 12 (1931), pp. 349ff.

Bence Jones, Henry. *The Life and Letters of Faraday* (2 vols.; London: Longmans, Green, and Co., 1870)

Berle, Adolf A. and G. C. Means. *The Modern Corporation and Private Property* (New York: Macmillan, 1982)

Brewster, David. *Memoirs of the Life, Writings, and Discoveries of Sir Isaac Newton*, 2 vols. (Edinburgh: T. Constable & Co., 1855)

Cajori, Florian, ed., *Newton's Principia, Motte's translation revised* (Berkeley, California: University of California Press, 1960)

Campbell, Lewis and William Garnett. *The Life of James Clerk Maxwell* (London: Macmillan and Co., 2nd ed., 1884)

Castillejo, David. *The Expanding Force in Newton's Cosmos* (Madrid: Ediciones de arte, 1981)

Cohen, I. B., ed., *Isaac Newton's Papers & Letters on Natural Philosophy*, 2nd edn. (Cambridge, Mass.: Harvard University Press, 1978)

Davie, George Elder. *The Democratic Intellect: Scotland and Her Universities in the Nineteenth Century* (Edinburgh: Edinburgh University Press, 1962)

Densmore, Dana. *Newton's Principia, The Central Argument* (3rd edn.; Santa Fe: Green Lion Press, 2003)

Descartes, René. *The Geometry of René Descartes* (Dover Publications, 1954)

_____. *Le Monde: ou, traité de la lumière*, trans. Michael Mahoney (New York: Abaris Books, 1979)

_____. *The Philosophical Works of Descartes*, trans. Elizabeth S. Haldane and G. R. T. Ross. New York: Dover Publications, 1955) Includes *Discourse on Method, Meditations, Rules for the Direction of the Mind*, and *Principles of Philosophy*.

Dobbs, Betty Jo Teeter. *The Foundations of Newton's Alchemy, or "The Hunting of the Greene Lyon"* (Cambridge, England: Cambridge University Press, 1975)

Duhem, Pierre. *Les théories électriques de J. Clerk Maxwell* (Paris: A. Hermann, 1902)

Euclid. *Elements, All Thirteen Books Complete in One Volume* (Santa Fe: Green Lion Press, 2010)

Faraday, Michael. *Experimental Researches in Electricity* (3 vols.; London: Taylor & Francis, 1839, 1844, 1855; reprinted Santa Fe: Green Lion Press, 2000) Short form reference: *XR*, followed by paragraph number.

Hamilton, William. *Lectures on Metaphysics and Logic* (2 vols.; Boston: 1859)

Hegel, G. W. F. *Philosophy of History* , tr. J. Sibree (1899; reprinted Dover Publications, 1956)

Hertz, Heinrich. *Electric Waves*, tr. D. E. Jones (London: Macmillan, 1893; reprinted Dover Publications, 1962)

Hobbes, Thomas. *Leviathan* (Penguin Books, 1968)

Horsley, Samuel. *Isaaci Newtoni Opera Omnia*, 5 vols. (1779–1785)

Kautsky, Karl. *Theories of Surplas Value* (Moscow: 1963–1971)

Koyré, Alexandre. *From the Closed World to the Infinite Universe* (New York: Harper, 1958; reprinted Peter Smith, 1983)

Locke, John. *Second Treatise on Civil Government* (Bobbs–Merrill, 1952)

Lucretius, *De rerum natura* (On the Nature of Things), tr. H. A. J. Munro (New York: Washington Square Press, 1965)

Lukács, Georg, "Reification and the Consciousness of the Proletariat," in *History and Class Consciousness* (Cambridge, Mass.: The MIT Press, 1971)

Manuel, Frank E. *Isaac Newton, Historian* (Cambridge, Mass.: Harvard University Press, 1963)

_____. *A Portrait of Isaac Newton* (Cambridge, Mass.: Harvard University Press, 1968)

_____. *The Religion of Isaac Newton* (Oxford: Clarendon Press, 1974)

Martin, Thomas, ed. *Faraday's Diary* (7 vols. and Index vol.; London: G. Bell & Sons, 1932–36)

Marx, Karl. *Capital*, tr. Samuel Moore and Edward Aveling (3 vols.; New York: International Publishers, 1967)
Also tr. David Fernbach (New York: Vintage Books, 1978, 1981)

_____. *Early Writings*, ed. T. B. Bottomore (New York: McGraw–Hill, 1964)

_____. *German Ideology* (New York: International Publishers, 1970)

Maxwell, James Clerk. *Matter and Motion* (Society for Promoting Christian Knowledge 1876). Reprinted Dover Publications (1991)

_____. *A Treatise on Electricity and Magnetism* (2 vols.; Oxford: Clarendon Press, 1873). Third edition, ed. J. J. Thomson (2 vols., 1892; reprinted Dover Publications, 1954)

McLachlan, Herbert, ed. *Sir Isaac Newton: Theological Manuscripts* (Liverpool: Liverpool University Press, 1950)

Mehring, Franz. *Karl Marx* (Ann Arbor: 1962)

More, Louis Trenchard. *Isaac Newton, a Biography* (New York: Dover Publications, 1934)

Newton, Isaac. *Opticks* (London, 1730; reprinted Dover Publications, 1952)

Niven, W. D., ed. *The Scientific Papers of James Clerk Maxwell* (2 vols.; Cambridge: Cambridge University Press, 1890; reprinted Dover Publications, 1952)

Parker, Richard. *The Myth of the Middle Class* (New York: Liveright, 1972)

Pepper, Jon. "Newton's Mathematical Work," in John Fauvel et al., eds., *Let Newton Be!* (New York: Oxford University Press, 1988)

Plato. *Dialogues of Plato*, tr. Benjamin Jowett (2 vols.; Random House, 1937). References to individual dialogues specify Stephanus line numbers.

Ptolemy. *Almagest*, tr. R. Catesby Taliaferro (Chicago: Encyclopædia Britannica, Inc., 1948)
Also tr. G. J. Toomer (Princeton University Press, 1998)

Rousseau, Jean–Jacques. *The Social Contract* and *Discourse on the Origin of Inequality* (Washington Square Press, 1967)

Smith, Adam. *Inquiry into the Nature and Causes of the Wealth of Nations* (Modern Library, 1965)

St. Augustine, *Confessions*; with *The City of God* and *De Doctrina Christiana* (On Christian Doctrine) (Chicago: Encyclopædia Britannica, Inc., 1952)

Staff, Social Sciences 1, The College of the University of Chicago. *The People Shall Judge* (2 vols.; Chicago: University of Chicago Press, 1949)

Syntopicon (*GBWW*, Vols. 2 and 3) (Chicago: Encyclopædia Britannica, Inc., 1952)

Thomson, William and P. G. Tait. *Treatise on Natural Philosophy* (Oxford: 1867)

Tyndall, John. *Faraday as a Discoverer* (Thomas Y. Crowell Co., 1961)

Westfall, Richard. *Force in Newton's Physics* (New York: American Elsevier, 1971)

_____. *Never at Rest* (Cambridge, England: Cambridge University Press, 1980)

Whiteside, D. T. "Sources and Strengths of Newton's Early Mathematical Papers," in Robert Paleter, ed., *The Annus Mirabilis of Sir Isaac Newton 1666–1966* (Cambridge, Mass.: The MIT Press, 1967)

Whyte, Lancelot, ed. *Roger Joseph Boscovich* (London: Allen and Unwin, 1961)

Williams, L. Pearce, ed. *The Selected Correspondence of Michael Faraday* (2 vols.; Cambridge: Cambridge University Press, 1971)

Wilson, Curtis. "Newton's Path to the *Principia*" (*GIT* 1985, pp. 179–229)

Index

About the Author

Thomas King Simpson is Tutor Emeritus at St. John's College in Annapolis, Maryland and Santa Fe, New Mexico. He has taught at the American University at Cairo and is a co–founder of The Key School in Annapolis, Maryland. He was educated at Rensselaer Polytechnic Institute, St. John's College, Wesleyan University, and the Johns Hopkins University, where in 1968 he received his doctorate in the history of science. His background also includes engineering and the classics. Simpson's wide range of interests extends to schools, museums, and in general towards broadening the role of what George Elder Davie has called "the democratic intellect."

Other books by Simpson include *Maxwell on the Electromagnetic Field* (Rutgers University Press, 1997), *Figures of Thought: A Literary Appreciation of Maxwell's* Treatise on Electricity and Magnetism (Green Lion Press, 2005), and *Maxwell's Mathematical Rhetoric* (Green Lion Press, 2010).